COVID-19 Lockdowns and the Urban Poor in Harare, Zimbabwe

Johannes Itai Bhanye • Fortune Mangara
Abraham R. Matamanda • Lameck Kachena

COVID-19 Lockdowns and the Urban Poor in Harare, Zimbabwe

Emerging Perspectives and the Morphing of a Sustainable Urban Future

Springer

Johannes Itai Bhanye
Department of Urban and Regional Planning
University of the Free State
Bloemfontein, South Africa

Abraham R. Matamanda
Department of Geography
University of Free State
Bloemfontein, South Africa

Fortune Mangara
African Centre for Disaster Studies
North-West University
Potchefstroom, South Africa

Lameck Kachena
Centre for Applied Social Sciences
University of Zimbabwe
Harare, Zimbabwe

ISBN 978-3-031-41668-2 ISBN 978-3-031-41669-9 (eBook)
https://doi.org/10.1007/978-3-031-41669-9

This Springer imprint is published by the registered company Springer Nature Switzerland AG
The registered company address is: Gewerbestrasse 11, 6330 Cham, Switzerland

Paper in this product is recyclable.

We dedicate this book to victims of the COVID-19 pandemic, families that lost their loved ones during this period, and those still grappling to recover and adjust from the throes of the pandemic.

Preface

The year 2020 brought an unprecedented global pandemic, which had not been seen in a century. The COVID-19 pandemic disrupted economies, social structures, and governance systems worldwide, leaving behind a trail of devastation that continues to be felt to this day. The pandemic, and the resultant lockdowns, had a profound impact on the most vulnerable populations, particularly the urban poor living in informal settlements or "slums." The idea of the book emerged from the need to give voice to the marginalized who often remain neglected and suffer the most during crises.

This book, *COVID-19 Lockdowns and the Urban Poor in Harare, Zimbabwe: Emerging Perspectives and the Morphing of a Sustainable Urban Future* focuses on the experiences of the urban poor in Harare, Zimbabwe, during the pandemic and the subsequent lockdowns. Drawing on the experiences of a diverse range of stakeholders, including residents, community leaders, civil society organizations, government officials, and secondary sources, this book examines the impact of COVID-19 on the lives of the urban poor. It explores potential pathways toward a more sustainable and resilient urban future.

The book emerges from a collaborative effort among scholars and researchers committed to understanding and addressing the challenges facing the urban poor in Harare, Zimbabwe, and Africa at large. The chapters in this book explore a range of issues that include the theoretical perspectives focusing on the urban penalty, complexity, distributive justice, and the right to the city theories in the context of the urban poor during the COVID-19 pandemic; contextualizing Harare urban, socio-economic profile and history of pandemics in the city; the COVID-19 pandemic and urban policy interventions in Zimbabwe; and exploring the socio-economic impacts of COVID-19 lockdowns on poor urbanites in Harare, Zimbabwe using remote ethnography. The final chapter concludes the book, exploring the changing landscape of urban sustainability in the wake of the pandemic.

As the world continues to grapple with the impacts of the COVID-19 in post-pandemic phase, the insights and perspectives presented in this book are timely and relevant. They offer important lessons for policymakers, practitioners, and academics to understand vulnerable urban populations' challenges and develop strategies for promoting a more sustainable and equitable urban future.

The book contributes to ongoing discussions and debates on the impacts of global pandemics on urban development, inspiring new ideas and approaches for addressing the complex challenges facing cities in the twenty-first century.

Bloemfontein, South Africa — Johannes Itai Bhanye
Potchefstroom, South Africa — Fortune Mangara
Bloemfontein, South Africa — Abraham R. Matamanda
Harare, Zimbabwe — Lameck Kachena

Acknowledgments

This book would not have been possible without the contributions of a range of individuals and organizations, all of whom played an important role in shaping its content and direction. We express our sincere gratitude to the following:

First and foremost, we would like to thank the residents of Harare, Zimbabwe, who generously shared their stories and experiences with us, physically and through remote ethnography. Your insights and perspectives have been invaluable in shaping this book, and we are deeply grateful for your time and commitment. We thank colleagues from the University of Zimbabwe, the University of the Free State, and other institutions who supported our research and contributed to understanding the challenges facing urban poor populations in Zimbabwe. Your expertise and guidance have been invaluable in shaping the content of this book. We also extend our appreciation to the non-governmental organizations and community-based organizations in Harare, which supported our research efforts and provided critical insights into the impact of COVID-19 on the urban poor.

Special thanks to Zachary Romano and the editorial team from Springer Nature for the unwavering support in this book project from the conceptualization phase right through to the publication. We also highly appreciate the critical comments and suggestions from the four anonymous reviewers, whose comments helped us refine the arguments and discussions presented in the book. The assistance with language and technical editing from Liesl van der Westhuizen from the University of the Free State is also greatly appreciated.

Finally, we express our gratitude to our families and loved ones, who supported us throughout the research and writing process. Your encouragement and understanding have been essential in helping us complete this book. Together, these individuals and organizations helped shape a book that we hope will make a meaningful

contribution to ongoing discussions and debates about the impact of COVID-19 on urban development, and the potential for sustainable and inclusive solutions to the challenges facing cities in the twenty-first century.
April 2023

Bloemfontein, South Africa Johannes Itai Bhanye
Potchefstroom, South Africa Fortune Mangara
Bloemfontein, South Africa Abraham R. Matamanda
Harare, Zimbabwe Lameck Kachena

Blurb About the Book

COVID-19 Lockdowns and the Urban Poor in Harare, Zimbabwe COVID-19: Emerging Perspectives and the Morphing of a Sustainable Urban Future is an insightful and timely book that provides a unique perspective on the impact of the COVID-19 pandemic on the urban poor in one of Africa's most populous cities, Harare. Across the world, governments imposed mandatory lockdowns as a containment measure against the devastating impacts of the COVID-19 pandemic. However, the lived experiences associated with mandatory lockdowns were experienced most by the most vulnerable in cities, who depend on the informal sector, without savings and safety nets. Whilst the COVID-19 pandemic presented new challenges to traditional fieldwork research, we innovatively embraced new and agile directions in research methodologies by adopting remote ethnography. We illuminate four major socio-economic impacts associated with COVID-19 lockdowns among the urban poor in Harare: The food security systems of poor urbanites who consume through the informal economy; closure of the informal sector; induced urban-rural migration; and strengthening of visible urban disparities between the poor and the rich. We conclude that the COVID-19 pandemic mandatory lockdowns deepened social and spatial inequality among the urban poor, threatening their right to the city; highlighting the urgent need for more sustainable and inclusive urban planning and development. We also argue that the pandemic presented unique urban challenges, exposing deficiencies of current urban governance models in ensuring the right to the city by all and realizing Sustainable Development Goals and the New Urban Agenda. However, the pandemic also gave planners and policymakers a unique chance to re-envision pandemic-resilient cities that are more just, equitable, resilient, and sustainable. We also find insights from this book useful to guide state and non-state social service providers to reorder social safety nets to accommodate marginalized and vulnerable urban populations during global catastrophes. The book offers valuable insights into the complex issues facing cities in the Global South, and the role that local communities and grassroots organizations can play in shaping a more equitable and resilient future. This thought-provoking and informative book is essential reading for anyone interested in the intersection of urban development, public health, and social justice in the wake of urban crises and pandemics like COVID-19.

Contents

About the Authors

Johannes Itai Bhanye is a researcher and academic in Migration and Urban Planning Studies, with strong academic credentials combined with "real-world" experience. Dr. Bhanye has conducted research projects in Southern Africa, including Zimbabwe, South Africa, and Lesotho. He is currently a Postdoctoral Research Fellow at the Department of Urban and Regional Planning, University of the Free State, in South Africa. Dr. Bhanye holds a Ph.D. in Migration and Land Settlement from the University of Zimbabwe, with fellowship support by The Andrew W. Mellon Foundation. His research interests encompass land governance, peri-urbanization, migration, urban informality, food systems, cities, and social change among other development-related topics in Africa. He has been invited to speak at several International Conferences and seminars in China, Switzerland, Belgium, Germany, South Africa, Lesotho, Morocco, Zambia, and Ghana.

Fortune Mangara holds a qualification in Rural and Urban Planning from the University of Zimbabwe and earned an M.Sc. and a Ph.D. in Urban and Regional Planning from the North-West University. Presently, he is a Postdoctoral Research Fellow affiliated with the African Centre for Disaster Studies at the School of Geo and Spatial Sciences located at the North-West University. Dr. Mangara's research focuses on spatial and urban resilience, disaster risk studies, disaster resilience, disaster risk governance, urban disaster risk, social and economic geography, urban policy, urban governance, transport planning and mobility, and spatial transformations in African cities.

Abraham R. Matamanda (Ph.D. Urban and Regional Planning) is an NRF Y2-rated Urban and Regional Planner who also trained as a social ecologist. Abraham lectures at the University of Free State (UFS) in the Department of Geography. Abraham is currently the editor of the *Town Planning Journal* published by UFS and also serves on the academic editorial board of *Plos Water Journal*. He is a Fellow of the Department of Higher Education and Training Future Professorate Programme Phase 1, third Cohort. Abraham is the South African PI for a global collaborative research project exploring how children and young people from monetary poor households have adapted to the COVID-19 pandemic focusing on food, education, and play/leisure (https://panexyouth.com/). His research focuses on urban governance and planning, climate change adaptation and resilience, informal Global South urbanism, urban food systems, and housing studies. Abraham is co-editor and co-author of *Urban Geography in Postcolonial Zimbabwe: Paradigms and Perspectives for Sustainable Urban Planning and Governance* and *Housing and Technology: A Special Focus on Zimbabwe* both published with Springer Nature in 2021.

Lameck Kachena is an early career social geographer with hands-on experience in socio-environmental interactions in borderland regions of Southeast Africa (Zimbabwe, Mozambique, and South Africa). Inspired by critical theory and co-production of knowledge, his current research is informed by ethnographic methods, creative methods, and art-based tools to understand insecurities, injustice, and inequalities associated with dominant discourse and policies on marginalized social groups. Mr. Kachena holds a Certificate in Climate Change and its Impacts (Brown University-USA), M.Sc. in Social Ecology, and B.Sc. in Sociology (University of Zimbabwe).

Acronyms

AIDS	Acquired Immunodeficiency Syndrome
CBD	Central Business District
COVID-19	Coronavirus Disease of 2019
DRC	Democratic Republic of Congo
HIV	Human Immunodeficiency Virus
ICT	Information Communication Technology
IFPRI	International Food Policy Research Institute
NUA	New Urban Agenda
OECD	Organisation for Economic Cooperation and Development
SADC	Southern African Development Community
TCV	Typhoid Conjugate Vaccine
UFS	University of the Free State
UN	United Nations
WHO	World Health Organization
ZANU-PF	Zimbabwe African National Union—Patriotic Front
ZNLWVA	Zimbabwe National Liberation War Veterans Association
ZSE	Zimbabwe Stock Exchange

Chapter 1
Introduction

Introduction

Pandemics have plagued and ravaged humanity since the earliest days (Dasgupta and Crunkhorn 2020; Hays 2005). The Antonine Plague (165 AD), the Plague of Justinian (541 AD), the Black Death (1350), the Great Plague of London (1665), the Spanish flu (1918), and the Asian flu (1957) have all caused severe damage to humans ranging from deaths to complete closure of economies (Huremović 2019). In the twenty-first century, the risk of disease outbreaks and their spread and amplification into pandemics have drastically increased due to rapid urbanization, travel and trade, and environmental disruptions (Huremović 2019; LePan 2020). In 2019, the novel coronavirus (COVID-19) was reported in China, with its origin traced to a food market in Wuhan City. At the time, the virus and infections were only reported in China. However, owing to globalization and porous international borders, the virus spread to other parts of the world within weeks of the first case detected in China (Bhanye and Bhanye 2023). Considering the rapid spread and mortality rates associated with COVID-19, the World Health Organization (WHO) declared COVID-19 a global pandemic in March 2020. Since then, life has never been the same as COVID-19 created short-term disruptions and provoked long-term changes in how the world lives and operates.

To contain the spread of COVID-19, mandatory lockdowns characterized by restricted movement and social distancing were imposed in many countries (Matsungo and Chopera 2020). Lockdown was a radical measure introduced to confine individuals and communities to limit movement and social interaction, thereby minimizing the spread of the highly contagious COVID-19. While the term lockdown became ubiquitous with controlling the spread of the pandemic, the term itself remained vague. Several variations of lockdowns were also introduced, including mandatory, compulsory, voluntary, and partial (Bhanye and Bhanye 2023; Nduna and Tshona 2021). Haider et al. (2020: 2) defined lockdown as "a set of measures

J. I. Bhanye et al., *COVID-19 Lockdowns and the Urban Poor in Harare, Zimbabwe*, https://doi.org/10.1007/978-3-031-41669-9_1

aimed at reducing transmission of COVID-19 that are mandatory, applied indiscriminately to a general population, and involve some restrictions on the established pattern of social and economic life." The definition brings to attention the impartiality of the measures that are applied to all individuals and groups in a particular area without prejudice. Being mandatory brings in policing and enforcement to ensure the restrictions are observed (Bhanye and Bhanye 2023; Bhanye et al. 2023). Hence, different measures and approaches were used, ranging from subtle to violent confrontations, as noted in some instances in Zimbabwe, South Africa, and India (Nduna and Tshona 2021). Moreover, lockdown is associated with restricted movement, confinement, and disruptions of normal socioeconomic activities.

The Government of Zimbabwe declared a state of national disaster in response to the COVID-19 pandemic on March 27, 2020, Friday (Mutanda 2022). This was followed by a strict nationwide lockdown on March 30, 2020, resulting in intercity travel bans, closure of educational institutions, workplaces, stadiums, theatres, and shopping centers and mandatory quarantines of potentially infected people (Tom and Chipenda 2020). While high-income countries like the United States of America, China, Germany, and the United Kingdom witnessed benefits of lockdowns as a containment strategy for COVID-19 through a reduction in hospital admissions and deaths, there were far-reaching negative impacts in developing countries (Teachout and Zipfel 2020).

The earliest studies on COVID-19 show that lockdowns led to much more significant and rapid contractions of economic activity than previous crises, including the global food crisis of 2007–2008 and the 2009 recession (Teachout and Zipfel 2020; Thurlow 2020). In developing countries, many urban poor depend on daily incomes from informal work; lockdown policies undermined such households' survival opportunities (Chirisa et al. 2020; Matamanda et al. 2022). Like many countries in sub-Saharan Africa, Zimbabwe has an under-resourced healthcare system, high unemployment, densely populated urban areas, and shortages of basic commodities (including water and food) (Matamanda and Nel 2021), making lockdowns difficult to adhere to and enforced on the urban poor (Mackworth-Young et al. 2021).

We conclude that the COVID-19 pandemic mandatory lockdowns deepened social and spatial inequality among the urban poor, threatening their right to the city. The socioeconomic impacts upsurged poverty and increased unemployment and the risks of hunger and food insecurity, reinforcing existing inequalities and breaking social harmony in cities, even past the COVID-19 pandemic. We argue that the pandemic presented unique urban challenges, exposing deficiencies of current urban governance models, ensuring the right to the city by all, and realizing Sustainable Development Goals, in particular Goal 11,[1] making cities inclusive, safe, resilient,

[1] Sustainable Development Goal (SDG) 11 is about making cities and human settlements inclusive, safe, resilient, and sustainable. Today, more than half the world's population live in cities. By 2050, an estimated 7 out of 10 people will likely live in urban areas. Achieving SDG 11 is critical for promoting social and economic development, enhancing public health, mitigating climate change, ensuring disaster resilience, fostering social inclusion, and supporting sustainable consumption and production.

and sustainable and the New Urban Agenda (NUA).[2] However, the pandemic also gives planners and policymakers a unique chance to re-envision pandemic-resilient cities that are more just, equitable, resilient, and sustainable. We also find insights from this study useful to guide state and non-state social service providers to reorder social safety nets to accommodate marginalized and vulnerable urban populations during global catastrophes.

Aim and Objective of the Book

This book focuses on the socioeconomic impacts of the COVID-19 pandemic and associated lockdowns on the welfare of the urban poor in Harare, Zimbabwe. While there is a growing number of studies on the impacts of COVID-19 mandatory lockdowns on the urban poor in Zimbabwe and across the globe, traditional research methodologies have been used, for example, structured questionnaires (Matsungo and Chopera 2020), document analysis (Chirisa et al. 2020), and qualitative face to face interviews (Dzawanda et al. 2021; Nyanga and Zirima 2020). However, the COVID-19 pandemic presents new challenges to traditional fieldwork research. In this book, we innovatively embraced new directions in research methodologies by adopting digital/remote ethnography[3] to deepen our understanding of how lockdown restrictions impacted the most vulnerable populations in urban settings.

Through remote ethnography, this book provides evidence-based stories of how COVID-19 mobility restrictions affected poor urbanites in Harare, Zimbabwe. Unlike traditional research approaches, this emerging and innovative methodological approach can nimbly cut across geographical boundaries and societal power structures. Remote ethnography also enables immediate communication between researchers and participants, providing an easy-to-use tool for data collection and various self-expression options, such as written text, photographs, and video recordings.

We start by presenting the theoretical underpinnings of the book focusing on the urban health penalty, the right to the city, complexity theory, and distributive justice

---

[2] The New Urban Agenda is a global agreement adopted by United Nations member states in 2016 at the United Nations Conference on Housing and Sustainable Urban Development (Habitat III). It is a comprehensive roadmap for sustainable urban development that guides policies and actions for cities and human settlements to support the Sustainable Development Goals. The New Urban Agenda recognizes the important role that cities play in driving economic growth, social inclusion, and environmental sustainability. It calls for an integrated and participatory approach to urban planning and development that involves all stakeholders, including local governments, civil society organizations, the private sector, and the general public.

[3] Remote ethnography (also known as "virtual ethnography," "cyber ethnography," "netnography," or "mobile ethnography") is a digital transformation of in-person ethnography and a form of online or remote ethnographic research. It leverages the power of technology—gadgets and the Internet—to help researchers generate contextual insights into the lived experiences of social groups remotely.

theory. These four theories situate the COVID-19 pandemic and its impacts on the urban poor in theoretical foundations, which raise issues of how the poor are affected by disease/health pandemics due to their living conditions. In the context of the book, we focus on the socioeconomic profiling of Harare, the history of pandemics in the city, and the vulnerabilities of the urban poor over time. The book also presents the Government of Zimbabwe's policy interventions to address the COVID-19 pandemic, demonstrating how the policies had little regard for the welfare of the urban poor, which resulted in the exacerbation of their suffering who struggled to make ends meet. In the main finding chapter, the book illuminates four major socioeconomic impacts associated with COVID-19 lockdowns among the urban poor: the food security systems of poor urbanites who access food through the informal economy; closure of the informal sector; induced urban–rural migration; and strengthening of visible urban disparities between the poor and the rich. The final chapter, *From Crisis to Action: Emerging Perspectives and the Morphing of a Sustainable Urban Future Post-COVID-19 Pandemic*, concludes the book, exploring the changing landscape of urban sustainability in the wake of the pandemic. The chapter presents how the pandemic affects urban systems, particularly the urban poor, and highlights innovative solutions and emerging perspectives that can shape a more sustainable and resilient future.

Structure of the Book

This book comprises six chapters that focus on the COVID-19 pandemic policy interventions and the socioeconomic impacts of these policies on the urban poor. The outline of the respective chapters are as follows:

Chapter 1 sets the tone of the book by introducing the COVID-19 pandemic and its impact on African cities. The focus is on the urban poor and the digital ethnography employed to gather information for this book. The aim and scope of the book are introduced in this chapter. The chapter provides an introduction and synopsis of the COVID-19 pandemic and how the global responses, especially lockdowns, affected the urban poor. The central argument and broad theme guiding this book are presented in this chapter, discussing policy interventions implemented to curb the COVID-19 pandemic, which disproportionately affected the urban poor through loss of economic activities, lack of inclusion and consideration of their needs in policy formulation, and exacerbating food insecurities. A justification is provided on the articulation and focus on the four major socioeconomic impacts associated with COVID-19 lockdowns among the urban poor: the food security systems of poor urbanites who access food through the informal economy, closure of the informal sector, induced urban–rural migration, and strengthening of visible urban disparities between the poor and the rich.

Chapter 2 presents the theoretical underpinnings of this book. The chapter is guided by four theories that show the nuances and intricacies of the COVID-19 pandemic and its impacts on the urban poor. These theories include the urban health penalty, the right to the city, complexity theory, and distributive justice theory. The theories are discussed, bringing out their connection to the socioeconomic impacts of the pandemic on the urban poor, who tend to be marginalized in decision-making and how the nature of the COVID-19 pandemic is highlighted through the complexity theory.

Chapter 3, *Contextualizing Harare Urban and the History of the Pandemic in the City*, profiles Harare as the case study context of this book. Specifically, the chapter addresses the socioeconomic and demographic profile of the city, taking a histological perspective. The chapter discusses the urbanization and migration patterns in Harare and how these influenced employment trends and urban livability in the city. What is apparent is the marginalization of blacks in Harare with segregated residential development during colonial times. Further, the chapter presents the cumulative breakdown of the political and economic spheres of the country resulting in the urbanization of poverty and rising informality. Finally, the chapter presents a brief history of epidemics/pandemics that have hit the city over time, specifically the urban poor, focusing on the decline of health infrastructure, pandemics including the cholera outbreak of 1992, HIV/AIDS pandemic in the 1990s, and the cholera and typhoid outbreak of 2008/2009. All had a substantial negative impact on the city's population, particularly the poor and those living in informal and emerging settlements.

Chapter 4, *COVID-19 and Urban Policy Interventions in Zimbabwe*, focuses on the Government of Zimbabwe's policy interventions to address the COVID-19 pandemic. The chapter acknowledges that the COVID-19 pandemic was a complex issue that required a concerted effort from different stakeholders to curb the impacts and reduce the vulnerabilities, especially among the urban poor. In this chapter, we define policies as any action the government decides to do or not do regarding a particular phenomenon. The chapter reveals that most policies were top-down and interventionistic as they were directives from the WHO, which dictated global interventions. Subsequently, these policies had little regard for the welfare of the urban poor resulting in the exacerbation of suffering who struggled to make ends meet.

In Chap. 5, Harare, the capital city of Zimbabwe, is used as a case study, which presents the socioeconomic impacts of COVID-19 mandatory lockdowns on poor urbanites. The chapter explains the digital ethnography used to gather data for the book. Also, it illuminates four major socioeconomic impacts associated with COVID-19 lockdowns among the urban poor: the food security systems of poor urbanites who access food through the informal economy, closure of the informal sector, induced urban–rural migration, and strengthening of visible urban disparities between the poor and the rich.

Lastly, Chap. 6 concludes the book. It summarizes the book and concludes the COVID-19 pandemic and related socioeconomic impacts on the urban poor. Future directions on addressing disease pandemics are articulated through lessons learnt from analyzing data on Harare.

References

Bhanye J, Bhanye A (2023) A blessing or a curse?: the role of social media during the COVID-19 pandemic in Africa. In: The Palgrave handbook of global social problems. Palgrave Macmillan, Cham. https://doi.org/10.1007/978-3-030-68127-2_367-1

Bhanye J, Shayamunda R, Tavirai RC (2023) Social media in the African context. A review study on benefits and pitfalls. In: The Palgrave handbook of global social problems. Palgrave Macmillan, Cham. https://doi.org/10.1007/978-3-030-68127-2_366-1

Chirisa I, Mutambisi T, Chivenge M, Mabaso E, Matamanda AR, Ncube R (2020) The urban penalty of COVID-19 lockdowns across the globe: manifestations and lessons for Anglophone sub-Saharan Africa. GeoJournal 87(2):815–828

Dasgupta S, Crunkhorn R (2020) A history of pandemics over the ages and the human cost. The Physician 6(2). https://doi.org/10.38192/1.6.2.1

Dzawanda B, Matsa M, Nicolau M (2021) Poverty on the rise: the impact of the COVID-19 lockdown on the informal sector of Gweru, Zimbabwe. Int Soc Sci J 71(S1):81–96

Haider N, Osman AY, Gadzekpo A et al (2020) Lockdown measures in response to COVID-19 in nine sub-Saharan African countries. BMJ Glob Health 5:e003319. https://doi.org/10.1136/bmjgh-2020-003319

Hays JN (2005) Epidemics and pandemics: their impacts on human history. Abc-clio

Huremović D (ed) (2019) Psychiatry of pandemics: a mental health response to infection outbreak. Springer, Cham

LePan N (2020) Visualizing the history of pandemics. Visual Capitalist 14:00060–00020

Mackworth-Young CR, Chingono R, Mavodza C, McHugh G, Tembo M, Chikwari CD et al (2021) Community perspectives on the COVID-19 response, Zimbabwe. Bull World Health Organ 99(2):85

Matamanda AR, Nel V (2021) (un)healthy cities: reflections on urban public health in poor neighbourhoods. In: Matamanda RA, Nel V, Chirisa I (eds) Urban geography in postcolonial Zimbabwe: paradigms and perspectives for sustainable urban planning and governance. Springer Nature, Cham, pp 93–118

Matamanda AR, Dunn M, Nel V (2022) Broken bridges over troubled waters: COVID-19 and the urban poor residing in Dinaweng informal settlement, Bloemfontein, South Africa. S Afr Geogr J 104(3):309–327

Matsungo TM, Chopera P (2020) Effect of the COVID-19-induced lockdown on nutrition, health and lifestyle patterns among adults in Zimbabwe. BMJ Nutr Prev Health 3(2):205

Mutanda D (2022) Challenges and opportunities for Zimbabwe's responses to COVID-19. Cogent Soc Sci 8(1):2084890

Nduna M, Tshona SO (2021) Domesticated poly-violence against women during the 2020 Covid-19 lockdown in South Africa. Psychol Stud 66(3):347–353

Nyanga T, Zirima H (2020) Reactions of small to medium enterprises in Masvingo, Zimbabwe to covid 19: implications on productivity. Bus Excell Manag 10(1):22–32

Teachout M, Zipfel C (2020) The economic impact of COVID-19 lockdowns in sub-Saharan Africa. Int Growth Centre 1(1):1–16

Thurlow J (2020) COVID-19 lockdowns have imposed substantial economic costs on countries in Africa, IFPRI book chapters. In: COVID-19 and global food security. International Food Policy Research Institute (IFPRI), pp 23–25

Tom T, Chipenda C (2020) COVID-19, lockdown and the family in Zimbabwe. J Comp Fam Stud 51(3–4):288–300

Chapter 2
Urban Penalty and the Right to the City of the Urban Poor During the COVID-19 Pandemic

Introduction

This chapter discusses the theoretical debates on how the urban poor are marginalized in cities and how they experience health inequalities. Since medieval times, cities have always been divided. The division in cities largely stems from income disparities and social classes that marginalized the poor (Morris 1994). The plight of the poor was apparent in industrial cities where urbanization rates were high and the living conditions among the urban poor were characterized by overcrowding (Mumford 1961). Hall (1996: 13) described the status quo as *the city of the dreadful night* due to the poor living conditions, including prostitution, urban violence, and lack of adequate urban services such as water and sanitation facilities. These appalling living conditions were evident in how the cholera epidemics that plagued Central London between 1848 and 1854 were concentrated around Broad Street in Soho, which had unhygienic industries and housing (Tulchinsky 2018). The discussion presented in the chapter focuses on the urban health penalty, the right to the city, complexity theory, and distributive justice theory.

First, urban health penalty theory explores the vulnerabilities that the poor face due to their positioning in the city's grey spaces, which are characterized by deprivations, abjection, and poverty, exposing them to health risks. These vulnerabilities are examined and interrogated through the lens of the COVID-19 pandemic. Second, the chapter situates the right to the city theory in context of the COVID-19 pandemic. Here, the focus is on government interventions and how such upheld or compromised the "citizens" rights to the city, especially the urban poor, who are often marginalized and excluded from accessing urban services and facilities. Third, the distributive justice theory is explored, highlighting how citizens are disadvantaged or benefited from accessing services in the city. Regarding COVID-19, this theory helps to understand how the poor benefited or were disadvantaged in accessing services being distributed to curb the spread of the pandemic or in recovering

J. I. Bhanye et al., *COVID-19 Lockdowns and the Urban Poor in Harare, Zimbabwe*, https://doi.org/10.1007/978-3-031-41669-9_2

from the pandemic. Moreover, the distributive justice theory also traces the pre-COVID-19 context and how the poor benefited or failed to benefit from services. Lastly, the complex theory explains the nature of COVID-19, which has been elusive and difficult to comprehend, and how it has affected different aspects of urban life beyond the health aspect of citizens.

Poverty and the Urban Health Penalty of the Poor During the COVID-19 Pandemic

The urban health penalty approach argues that the urban poor are exposed to multiple health challenges due to unhealthy physical and social environments (Pacione 2009). Brawer et al. (2000: 1) explained that poverty zones accommodating poor residents are characterized by "higher numbers of minority groups, epicenters for economic decline, abandoned homes, low wages, single-parent households, poor public schools, job loss, violence, hopelessness, and homelessness." These multiple socioeconomic and physical deprivations expose residents to health vulnerabilities that result in the "urban health penalty." Urban ills characterize these poverty zones as the poor often lack access to basic services such as water and sanitation (Bhanye 2022). This deprivation is common in many African cities, where residents are left to improvise water and sanitation services (Bhanye and Dzingirai 2020). Sewerage streams are common in most poor neighborhoods where residents are exposed to stench odors while risking waterborne disease. The flowing raw sewage and uncollected solid waste make them vulnerable to several infectious diseases (Fig. 2.1).

With the rapid urbanization of African cities, poverty levels have soared. Potts (1995: 248) demonstrated how the emergence of "a new urban poor" was experienced in African cities in the 1980s. This group of urban poor people comprised highly vulnerable individuals and communities, which constituted most of the urban population in postcolonial African cities such as Lusaka in Zambia, Harare in Zimbabwe, Accra in Ghana, and Lagos in Nigeria, among others (Bekker and Therborn 2012). The result of the proliferation of this new urban poor in cities is what Huchzermeyer (2011) described as "cities with slums." In her description, Huchzermeyer (2011: 25) pointed out that:

> the post-independence context African cities are marked by several crises, which explain in part the growing economic 'informalization', the backlog in formal or adequate housing and the proliferation of largely unprotected living arrangements.

Due to "relaxed" migration laws in postcolonial African cities, there has been an exodus of rural–urban migration as people rushed to the cities, previously enclaves of the Europeans. Rural push factors have primarily driven migration, for example, rural poverty in Zimbabwe, lack of economic opportunities, and a need for better living conditions (Andersson Djurfeldt 2012; Mutambirwa and Potts 1990; Wekwete 1988). Other factors, including environmental challenges

Fig. 2.1 Uncollected waste heap in an informal settlement in Bloemfontein, South Africa. (Source: Authors 2021)

such as climate change, political instability, and global economic crises, have exacerbated urban poverty in African cities.

With the increasing urban population, there has been an increase in demand for urban services, which has been difficult for the African governments to meet. Against the background of a colonial legacy of socio-spatial segregation, the demand for urban services exceeds the supply resulting in huge service delivery deficits, which threaten urban sustainability. This persisting backlog in formal and adequate housing is evident in most southern African cities where housing backlogs are high (Chirisa and Matamanda 2016). In South Africa, 2.5 million families were estimated to require housing in 2022, and the numbers kept increasing due to inadequate policy and financial capacity to address this problem (Houston 2022). For Zimbabwe, the housing backlog was estimated to be around 1.25 million in 2015 (Government of Zimbabwe 2015), showing a stark contrast between housing demand and supply as the government could only deliver a few housing units over the years (Matamanda et al. 2022c). Housing finance is limited and often beyond the reach of the poor. Soaring unemployment, the inability of national and local governments to provide serviced land and infrastructure to ever-growing urban populations, and politics of difference result in marginalization of certain individuals and groups in societies (Bah et al. 2018; Myers 2011; Bekker and Therborn 2012).

Fig. 2.2 Overcrowding is depicted through housing informally developed in Hopley Farm Settlement, Harare, Zimbabwe

Inadequate housing characterized by overcrowding (see Fig. 2.2), lack of reticulated water and sewer facilities, and high unemployment akin to the industrial cities where squalor was prevalent is common in many African cities (Nyashanu et al. 2020). Wanjiru and Matsubara (2017) highlighted that 135 informal settlements cover approximately 1% of Nairobi's total area while accommodating approximately half of the city's population. In Zimbabwe, settlements such as Hopley Farm in Harare, established in 2005, are overcrowded, making the spread of infectious diseases easier (see Fig. 2.2). Residents live in abject poverty and are exposed to multiple health challenges, including violence, sexually transmitted diseases, and waterborne diseases such as cholera and typhoid (Bhanye 2023a; Matamanda 2020). Maternal health problems are also experienced in this poor neighborhood due to the inability of residents to access healthcare facilities (Benhura 2016). In other African cities, poor neighborhoods are disconnected from the urban core; thus, residents lack access to health facilities. For example, residents in Caleb Motshabi settlement in Bloemfontein, South Africa, are located in an area where the roads are poor and this jeopardizes emergency services, while the nearest health facilities are not in easy reach (Matamanda et al. 2022a; Ntema et al. 2021).

Urban poverty is a complex phenomenon. Being poor in cities is associated with deprivations of essential goods and services necessary to support and maintain human well-being and good living standards. Among the essentials are physical goods and services such as adequate food, housing, potable water, reticulated sanitation facilities, and assets that enhance human well-being. Food poverty is a critical issue that is experienced in African cities. In Zimbabwe, a myriad of factors, including the persistent droughts since the early 2000s, political unrest, and impacts of the Fast Track Land Reform Programme, have exacerbated urban food security, revealing the magnitude of urban food poverty. The situation is dire in poor neighborhoods such as Epworth, Mabvuku, Hopley, and Tafara (Manjengwa et al. 2016).

Food insecurity is associated with health challenges. For children, malnutrition compromises their immune system, thus making them vulnerable to disease.

Urban health inequalities are aggravated by urban poverty as the poor lack health insurance, reside in marginal spaces lacking reticulated water and sewerage facilities, and are exposed to insecurity and violence (Matamanda and Nel 2021). Decision-making is critical in urban development, and the poor cannot make decisions on matters of well-being. The same applies to their lack of power and voice to make decisions in times of crises. Decisions tend to be top-down without understanding the implications for the poor. The result tends to be defiance as the poor fail to take heed of certain communications, even when the government's intentions may be noble (Stivers 2007). For example, during the aftermath of Hurricane Katrina, in New Orleans, United States of America, the poor failed to take heed of official communication. The result was frustrated risk response efforts due to the inappropriate communication used by the officials.

Right to the City Theory

Understanding the logic of the "right to the city" in context requires a brief discussion of the concept of human rights, which guides human activities. Human rights refer to norms and values that govern how individuals and groups in society should be treated and what is good for them to attain a good life. According to Marks (2016: 1):

> Human rights constitute a set of norms governing the treatment of individuals and groups by states and non-states actors based on ethical principles regarding what society considers fundamental to a decent life.

The quote brings to attention critical issues related to the right to the city, as shall be explained. First, rights are defined and regulated through institutions (local, regional, national, or international) that identify and respect norms and values, which inform how individuals and groups should be treated. Here the rationale is that there is a particular way of doing things, behaving, and relating to oneself and with others. The United Nations Charter of 1945 is critical in upholding human rights as it determined to "reaffirm faith in fundamental human rights, in the dignity and worth of the human person, in the equal rights of men and women and of nations large and small" (United Nations 1945: 2). Therefore, anything outside these norms espoused by the state and non-state actors is considered irrational and calls for sanctioning those transgressing this norm. Foucault (1981) queries this when he asks, "Who defines the norm?"

In light of the COVID-19 pandemic, the treatment of individuals and groups is linked to human rights, as many human rights violations occurred (Bhanye and Bhanye 2023). Cases were reported where the military and police were used to enforce national lockdowns and how they beat civilians found walking outside in defiance of lockdowns and curfews (Feltham 2020; UNAIDS 2020). A case to note was the death of Collins Khosa, a South African who died on April 10, 2020, in Alexandra Township (a poor neighborhood in Johannesburg). It is alleged that the late Collins Khosa was physically assaulted by members of the South African National Defence Force, who patrolled Alexandra Township during the country's hard lockdown (Makinana 2020). Commenting on this unfortunate incident, Judge Hans Fabricius ruled that this incident was a clear violation of Khoza's human rights and dignity, as he mentioned:

> rights to life, freedom and security, and treatment with dignity and respect, were among those provided for in South African law that may not be derogated from even in a state of emergency (Schatz 2021: online).

It was noted that certain groups were targeted, such as those operating in the informal sector (UNDP South Africa 2020). In some countries, citizens had to stay indoors and those found on the streets were usually reprimanded, as illustrated in the following quotation from Zimbabwe:

> However not everyone has been as lucky as this couple, as hundreds of other Zimbabweans can no longer leave their homes because of serious injuries inflicted on them by the security forces after gnawing hunger forced them to venture out. Those unlucky to be caught have harrowing stories to tell, having been savagely beaten by the armed soldiers and policemen patrolling business centres and residential area (Zenda 2020: online).

From this quote, it is evident how police and military brutality was prevalent to the extent that people feared leaving their homes to access basic services, thus confining them indoors. Such actions point to the violation of citizens to access the city, which in this instance was commercial facilities, while the ability to exercise was never tolerated.

Specific individuals were also targeted, for example, those operating in the informal sector. Governments undertook blitzes, thus compromising their human rights by threatening their social well-being and ethical values with a loss of dignity (Matamanda et al. 2022b; Parker et al. 2020). In Zimbabwe, such practices contradicts Section 51 of the Constitution of Zimbabwe, which upholds the right to human dignity (Government of Zimbabwe 2013).

There is an element of authority associated with rights. In this instance, the power issue comes to light as specific individuals determine the norm, as articulated by Foucault (1981). This issue is evident where institutions formulate laws and policies, which eventually become binding. Marks (2016) recognizes that these institutions include state and non-state actors. Most importantly, rights must uphold ethical standards that create a decent life among individuals and groups.

Human rights uphold the dignity and well-being of individuals and groups. This dignity resonates with the need for citizens to enjoy a decent life. A decent life in cities includes the ability of citizens to reside in less life-threatening situations and living conditions. Access to healthcare is considered a fundamental human right and is recognized in Section 76 of the Constitution of Zimbabwe (Government of Zimbabwe 2013).

Conceptualizing the City Through the Spatial Triad

The right to the city focuses on space. Henri Lefebvre postulated the theory of the right to the city in 1967. Lefebvre criticized the failure to include inhabitants in the production of urban spaces. In his first inception of the concept, Lefebvre emphasized the effects of capitalism on "the city," whereby urban life was downgraded into a commodity where social interaction became increasingly uprooted and urban space and governance were turned into exclusive goods (Lefebvre 1996). Central to the theory is how urbanites access and interact with urban space daily. In making sense of space, Lefebvre (1991: 38) argued three perspectives, which he termed "spatial triad." First, Lefebvre (1991:38) defined "perceived space" as:

> The spatial practice of a society secretes that space; it propounds and presupposes it in a dialectical interaction; it produces it slowly and surely as it masters and appropriates it. From the analytical standpoint, the spatial practice of a society is revealed through deciphering its space.

Here space is defined through social practices that give meaning to how individuals and communities use and value their surroundings. Of note is how social values and norms become intrinsic in shaping and giving meaning to space.

Second, conceived space was identified as abstract and symbolic. This refers to how experts conceive of space and includes planning documents, which show how experts envisage space. These experts often hold the power to shape space, thus bringing in codes to define space. Lefebvre (1991: 38) indicates that conceived space is:

> Space of scientists, planners, urbanists, technocratic subdividers and social engineers, as of a certain type of artist with a scientific bent- all of whom identify what is lived and perceived with what is conceived.

Unlike perceived space, which focuses on societal values and norms, conceived space is about the technocrats and how they make sense of space. These are often stakeholders who do not interact with the space daily but can imagine and reimagine the lived and conceived space, and Foucault (1981) defines them as individuals who define the norm. Through spatial planning and development control, the ombuds determine how space should be used, what it means to the users, and, thus, how it influences daily activities (see Fig. 2.3). For example, during the colonial and apartheid era, African cities were planned, and Africans were forced into townships that were planned and developed without the input of the locals, thus failing to integrate African values (Mphambukeli 2019). Mbembe (2019: 80) states that "space was thus the raw material of sovereignty and of the violence, it bears within it. Sovereignty meant occupation, and occupation meant relegating the colonized to a third zone between subjecthood and objecthood." In this regard, Lefebvre (1991) criticized the failure to include inhabitants in the production of urban spaces.

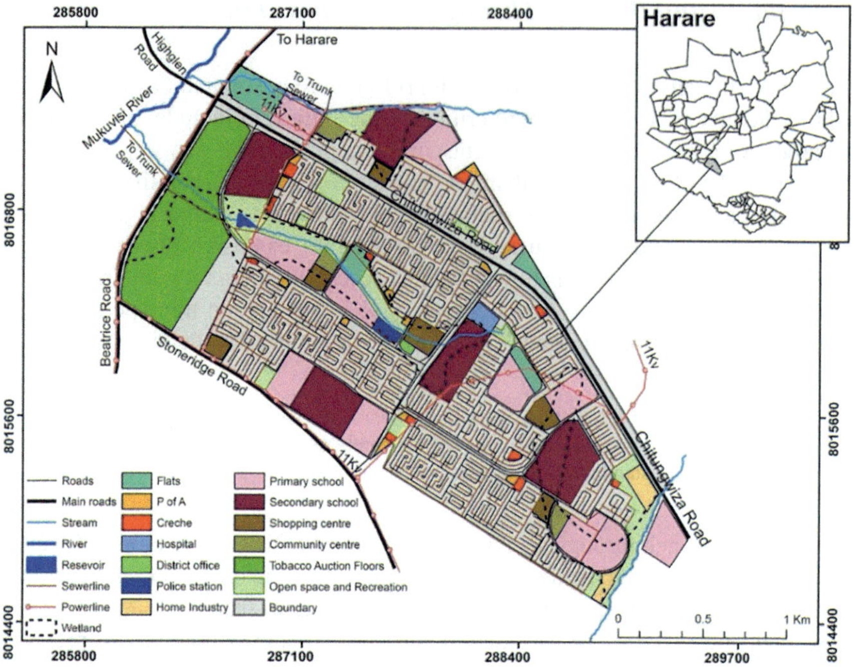

Fig. 2.3 A land-use planning scheme for Hopley Farm settlement, Harare, Zimbabwe. The scheme shows the envisaged land-use zoning, which planners prepared for the City of Harare. (Source: *Adopted from City of Harare*)

Lefebvre emphasized that urban life was downgraded into a commodity. In opposition to this trend, Lefebvre raised a call to "rescue the citizen as the main element and protagonist of the city that he had built" and to transform urban space into "a meeting point for building collective life" (Lefebvre 1996).

Third, Lefebvre (1991: 38) mentioned lived space as another dimension of space and postulated that it (lived space):

> Is space directly lived through its associated images and symbols, and hence the space of "inhabitants" and "users," but also of some artists and perhaps of those, such as a few writers and philosophers, who describe and aspire to do no more than describe. This is the dominated—and hence passively experienced—space which the imagination seeks to change and appropriate.

Therefore, this lived space is the actual physical space that individuals occupy and engage with (see Fig. 2.4). The houses people occupy become the lived space, as do the marketplaces that traders occupy and use while trading (see Fig. 2.5). The street vendors' streets also depict the lived spaces as humans interact physically with these spaces.

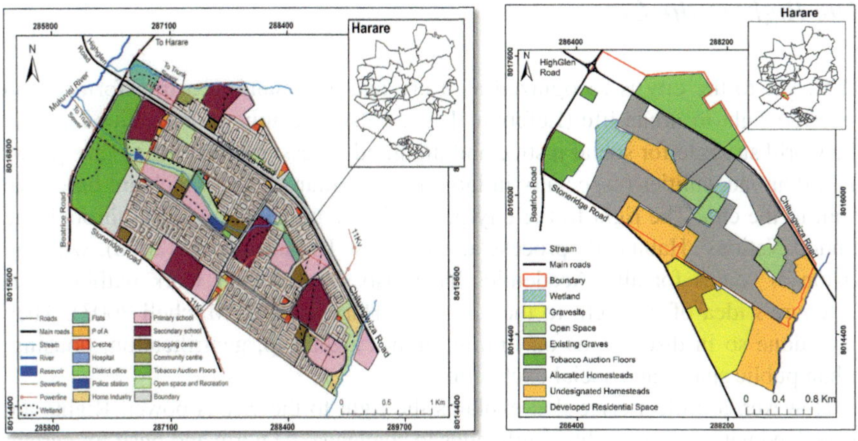

Fig. 2.4 The image on the left depicts the perceived space envisaged by planners from the City of Harare who prepared the land-use scheme for Hopley. The image on the right illustrates the lived space in Hopley, showing how the settlement was occupied, contradicting the envisaged plans

Fig. 2.5 On the left, informal traders occupy an undesignated place at Chikwanha Food Market, Chitungwiza, Zimbabwe. The image on the right shows women selling fruit and vegetables on a street corner in Maseru, Lesotho. (Source: Abraham R. Matamanda 2021)

The concept of the space triad contributes to understanding the nexus of the COVID-19 pandemic to the right to the city theory. The lockdowns were enforced, restricting people from accessing certain parts of the city. Likewise, the curfews were a means to time when certain spaces could be used. For example, during the lockdown in Zimbabwe, businesses operating in the central business district (CBD) had to open at 08:00 and close at 15:00. The CBD was thus a no-go area outside these times. In the residential areas, no one was allowed to roam the streets during the 22:00 to 05:30 curfew (Government of Zimbabwe 2022).

The Right to the City

The right to the city is recognized as a cry and a demand to a transformed and renewed right to urban life (Lefebvre 1991: 158). Recently, mobilizations around the world struggle for social justice and dignified access to urban life to face growing urban inequalities (especially in large metropolitan areas), which are based the right to the city. The right to the city gained international recognition through the United Nations' Habitat III process and the New Urban Agenda (2016), with the vision of "cities for all." Academics have also situated their work within Henri Lefebvre's idea of the right to the city (e.g., Harvey 2008; Mitchell 2003). They have done so in discussions on immigration, citizenship, gentrification, housing, urban public space, and social exclusion.

However, a critical issue that explains the right to the city is power. Rights are associated with power, enabling urbanites to challenge the norm and push for certain agendas that resonate with their needs. As Lefebvre (1991) articulated in the production of space and through the spatial triad, technocrats who create and recreate urban space often define the norm and, at times, have the power to influence the use of and access to urban space. This ability to create spaces can exclude certain individuals and groups from accessing certain parts of cities, a situation that eventually marginalizes some urbanites. The placement of infrastructure and services may also be a way in which some neighborhoods are segregated. Yiftachel (2009) noted that poor neighborhoods are neglected and kept in the shadows of the formal city. In this regard, the poor are thus confined in "camps" where they lack a voice and cannot access the privileges of the city (Kamete 2017). In such instances, power is presumed to be in the hands of the capitalist who control the means of production, hence defining the norm as to who benefits and suffers in the city. Capitalists are driven by profit. Therefore, the poor are least prioritized because they do not contribute much to profit, even though they may pay more for services they do not have than the elites.

Harvey (2012) and Mitchell (2003) argued that power is not only held by the capitalists who control the production of urban space. Instead, power can be managed democratically. Lefebvre (2003) mentions "interested parties" in the revolution for claiming the right to the city. These interested parties are not only capitalists but individuals and groups who seek to "de-alienate" urban space (Purcell 2014). Purcell (2002) argues that the right to the city is about influencing the appropriation of physical space and determining its use. Therefore, the right to the city includes ordinary citizens who advocate for the use and access of urban space to enable them to enjoy the city (Bhanye 2023b). However, the struggles by the poor, often through social movements, are deemed insignificant due to their inability to effect change. Referring to social movements, Harvey (2012: 120) remarks:

> If social movements are considered at all, they are typically constructed as either mere offshoots or displacements of these more fundamental struggles. Within the Marxist tradition, for example, urban struggles tend to be either ignored or dismissed as devoid of revolutionary potential or significance. Such struggles are constructed as being either about issues of reproduction rather than production or about rights, sovereignty, and citizenship, and therefore not about class.

There is not much detail about the threats of global pandemics on the urban poor through a right to the city perspective. Through detailed remote ethnographic lenses and an understanding of the city from below, we utilized the right to the city theory to describe how COVID-19 mandatory lockdowns perpetuated inequality and justice dominating the world's cities.

Distributive Justice Theory and the COVID-19 Pandemic

Justice is an elusive concept articulated from different perspectives in urban contexts. Building on the previous sections, we situate the debate in the context of space, where we focus on spatial justice as articulated by Soja (2010). Soja (2010: 62) argued that spatial (in) justice represents both process and outcome, as geographies or distributional patterns, which are in themselves just or unjust. These include the democratic processes undertaken by political and economic elites to benefit urban space for certain individuals and groups. In contrast, others are disadvantaged and treated as subjects and objects (Mbembe 2019: 80), thus perpetuating inequality in cities.

According to Soja (2009: 3), spatial injustices in cities are evident in locational discrimination resulting from the "biases imposed on certain populations because of their geographical location." Locational discrimination is exemplified in how the poor are disadvantaged in cities through being positioned in marginal and precarious spaces such as informal settlements on the city edges where they are deprived. Yiftachel (2009) termed these spaces "grey spaces" where opportunities are limited and the city authorities neglect the poor. The neglect is described by Mbembe (2019) as deliberate and the poor are plunged into abject spaces where they are meant to die slowly due to the vulnerability of the space.

Spatial injustice is also attributed to the political organization of space due to redlining of urban investments, exclusionary zoning of territorial apartheid, institutionalized residential segregation, and the imprint of colonial social control of space use (Soja 2009). Marginalization is often perpetuated against specific individuals and groups. For example, in Africa during colonial times, cities were the enclaves of European settlers who enjoyed more privileges and rights, while local Africans were disadvantaged in accessing services and faced deprivations in cities, which were defined along racial lines. Mbembe (2019: 80) explained that:

> Colonial occupation itself comprised seizing, delimiting, and asserting control over a geographical area – writing a new set of social and spatial relations on the ground. The writing of new spatial relations (territorialization) ultimately amounted to the production of boundaries and hierarchies, zones and enclaves; the subversion of existing property arrangements; the differential classification of people; resource extraction; and, finally, the manufacturing of a large reservoir of cultural imaginaries.

In contemporary African cities, spatial privileges and advantages are defined along class, as the poor suffer most of the deprivations in cities (Schindler 2017; Bozzoli 2019). Exclusionary zoning is evident in how the technocrats determine permitted land uses, which restrict certain groups from accessing the urban space. As argued by Lefebvre (1991) in his spatial triad, the technocrats are responsible for apportioning space and hence influence spatial outcomes when these decisions fail to cater to particular individuals.

Spatial injustice is explained through discontent. According to Marcuse (2010), discontent results from unequal treatment and forced constraint of specific people in cities, so they fail to explore unrestrained possibilities. Harvey (2012) identified foreign migrants as being marginalized and facing spatial injustices in cities. This is evident in South Africa, where xenophobic attacks targeted migrants who suffered injustices in South African cities with no access to certain rights and privileges. Political affiliation in cities is a critical issue that triggers unequal treatment. Young (1990) terms it politics of difference, which explains how some people are mistreated in cities due to political affiliation. The issues are apparent in times of crisis where aid is distributed along political lines.

Unequal distributive outcomes of capitalist urbanization represent spatial injustice. As alluded to in the previous section, cities are economic engines driven by capitalists' interests. However, the outcomes of capitalist urbanization are not always equally distributed as urban production or reproduction is channeled where profit is realized, resulting in injustices where certain areas are neglected and thus urbanites deprived of services and facilities.

Conclusion

This chapter discussed the theoretical underpinnings of the book. It highlights the spatial marginalization evident in how the poor are subjected to multiple health challenges due to the vulnerable spaces they reside in. As noted, the emerging health challenges point to the urban health penalty, amplified during disaster and crises. The poor are forced to reside in marginal and grey spaces due to socioeconomic challenges, political decisions, and actions that segregate individuals and groups along racial and income lines. Through zoning laws and urban planning ideologies, certain groups are confined in particular spaces, which are not necessarily habitable, thus compromising their right to the city. The injustices that emanate from these issues are highlighted, showing how the poor remain vulnerable. This chapter provides a theoretical background and context for Chap. 3, which explores the historical context of the disease pandemic and disasters in urban Zimbabwe.

References

Andersson Djurfeldt A (2012) Virtuous and vicious cycles in rural-urban linkages: cases from Zimbabwe. Africa Rev 4(2):136–156

Bah ME, Faye I, Geh ZF (2018) Housing market dynamics in Africa. Palgrave Macmillan, London

Bekker S, Therborn G (2012) Power and powerlessness: capital cities in Africa. HSRC Press, Cape Town
Benhura A (2016) Displaced and dispossessed: sexual and reproductive health rights for women in Hopley, Zimbabwe. Orient Anthropol 16(2):215–227
Bhanye J (2022) "Lydiate is now our home of a sort": perceptions of place amongst ageing first-generation Malawian migrants in Zimbabwe. Anthropol South Afr 45(3):180–194
Bhanye J (2023a) "Emerging forms of authority in land access?": the occult and witchcraft among Malawian migrants in peri-urban Zimbabwe. In: Debating religion and forced migration entanglements. Springer International Publishing, Cham, pp 89–107
Bhanye J (2023b) 'Emerging forms of spatialised and socialised authority' among 'tenure-insecure peri-urbanites' in African peri-urban spaces: a review study. Pan-Afr Convers 1(2):26–63
Bhanye J, Bhanye A (2023) A blessing or a curse?: the role of social media during the COVID-19 pandemic in Africa. In: The Palgrave handbook of global social problems. Palgrave Macmillan, Cham. https://doi.org/10.1007/978-3-030-68127-2_367-1
Bhanye J, Dzingirai V (2020) Plural strategies of accessing land among peri-urban squatters. Afr Black Diaspora Int J 13(1):98–113
Bozzoli B (2019) Theatres of struggle and the end of apartheid. Edinburgh University Press, Edinburgh
Brawer R, Plumb J, Carson-Weinstein L (2000) The urban health penalty and Jefferson Health System's community health collaboration. Health Policy Newsl 13(2):Article 6. Retrieved [29 December 2022] from http://jdc.jefferson.edu/hpn/vol13/iss2/6
Chirisa I, Matamanda AR (2016) Addressing urban poverty in Africa in the post-2015 period: perspectives for adequate and sustainable housing. J Settl Spat Plan 7(1):79–87
Feltham L (2020) Police and military abuses raise concerns amid lockdown defiance. Mail & Guardian, 30 Mar. https://mg.co.za/coronavirus-essentials/2020-03-30-police-and-military-abuses-raise-concerns-amid-lockdown-defiance/. Accessed 3 Dec 2022
Foucault M (1981) The order of discourse. In: Young RJC (ed) Untying the text: a post-structuralist reader. Routledge and Kegan Paul, London, pp 48–78
Government of Zimbabwe (2013) Constitution of Zimbabwe amendment No. 20. Government of Zimbabwe, Harare
Government of Zimbabwe (2015) National report for Habitat III. Government of Zimbabwe, Harare
Government of Zimbabwe (2022) Statutory Instrument 18 of 2022: public health (COVID-19 prevention, containment and treatment) (national lockdown) (No. 2) (amendment), order, 2022, No. 40. Government of Zimbabwe, Harare
Hall (1996) Cities of tomorrow: an intellectual history of urban planning design in the twentieth century. Wiley-Blackwell, Oxford
Harvey D (2008) The right to the city. City Read 6(1):23–40
Harvey D (2012) Rebel cities. From the right to the city to the urban revolution. Verso, London
Houston A (2022) Housing: time to replace the outdated one-plot-one-house model. We have to build much denser cities. GroundUp, 11 Oct. https://www.groundup.org.za/article/housing-we-need-leaders-who-face-up-to-the-truth/#:~:text=With%20every%20passing%20year%2C%20South,increasing%20in%20frequency%20and%20violence. Accessed 10 Jan 2023
Huchzermeyer M (2011) Cities with 'slums': from informal settlement eradication to a right to the city in Africa. UCT Press, Claremont
Kamete AY (2017) Governing enclaves of informality: unscrambling the logic of the camp in urban Zimbabwe. Geoforum 81:76–86
Lefebvre H (1991) The production of space. (Translation by Donald Nicholson-Smith). Blackwell, Oxford
Lefebvre H (1996) Writings on cities. Blackwell, Oxford
Lefebvre H (2003) The urban revolution. University of Minnesota Press, Minneapolis
Makinana A (2020) Collins Khosa murder: military ombud finds that soldiers acted improperly. Sunday Times, 19 Aug. https://www.timeslive.co.za/politics/2020-08-19-collins-khosa-murder-military-ombud-finds-that-soldiers-acted-improperly/. Accessed 9 Jan 2023

Manjengwa J, Matema C, Tirivanhu D (2016) Understanding urban poverty in two high-density suburbs of Harare, Zimbabwe. Dev South Afr 33(1):23–38
Marcuse P (2010) Spatial justice: derivative but causal or social justice. In: Bret B, Gervais-Lambony P, Hancock C (eds) Justices et injustices spatiales. Presses Universitaires de Paris Quest, Paris, pp 76–92
Marks SP (2016) Human rights: a brief introduction. Harvard University, London
Matamanda AR (2020) Living in an emerging settlement: the story of Hopley farm settlement, Harare Zimbabwe. Urban Forum 31(4):473–487
Matamanda AR, Nel V (2021) (Un) healthy cities: reflections on urban public health in poor neighbourhoods. In: Matamanda AR, Nel V, Chirisa I (eds) Urban geography in postcolonial Zimbabwe. Springer Nature, Cham, pp 93–118
Matamanda AR, Dunn M, Nel V (2022a) Broken bridges over troubled waters: COVID-19 and the urban poor residing in Dinaweng informal settlement, Bloemfontein, South Africa. S Afr Geograph J 104(3):309–327
Matamanda AR, Nel V, Chanza N, Leboto-Khetsi L, Mangara F, Paradza P (2022b) The political economy of COVID-19 pandemic: lessons learned from the responses of local government in sub-Saharan Africa. In: Nunes Silva C (ed) Local government and the COVID-19 pandemic. Springer Nature, Cham, pp 103–128
Matamanda AR, Chirisa I, Rammile S, Marais M (2022c) Housing and technology: special focus on Zimbabwe, vol 37. Springer Nature, Cham
Mbembe A (2019) Necropolitics (translated by Corcoran S). Duke University Press, Durham and London
Mitchell D (2003) The right to the city: social justice and the fight for public space. Guilford Press, New York
Morris AEJ (1994) History of urban form: before the industrial revolutions. Prentice Hall, London
Mphambukeli T (2019) Apartheid. In: Orum AM (ed) The Wiley Blackwell encyclopedia of urban and regional studies. Wiley, New York, pp 1–6
Mumford L (1961) The city in history: its origins, its transformations, and its prospects, vol 67. Houghton Mifflin Harcourt
Mutambirwa CC, Potts D (1990) Changing patterns of African rural-urban migration and urbanization in Zimbabwe. East S Afr Geograph J 1(1):26–39
Myers G (2011) African cities: alternative visions of urban theory and practice. Zed Books, London
Ntema J, Anderson I, Marais L (2021) Housing and possible health implications in upgraded informal settlements: evidence from Mangaung township, South Africa. In: Housing and SDGs in urban Africa. Springer, Singapore, pp 71–85
Nyashanu M, Pfende F, Ekpenyong M (2020) Exploring the challenges faced by frontline workers in health and social care amid the COVID-19 pandemic: experiences of frontline workers in the English Midlands region, UK. J Interprof Care 34(5):655–661
Pacione M (2009) Urban geography: a global perspective. Routledge, London
Parker M, MacGregor H, Akello G (2020) COVID-19, public authority and enforcement. Med Anthropol 39(8):666–670
Potts D (1995) Shall we go home? Increasing urban poverty in African cities and migration processes. Geogr J 161(3):245–264
Purcell M (2002) Excavating Lefebvre: the right to the city and its urban politics of the inhabitant. GeoJournal 58:99–108
Purcell M (2014) Possible worlds: Henri Lefebvre and the right to the city. J Urban Aff 36(1):141–154
Schatz C (2021) Documentary|Slain by soldiers: the Collins Khosa Story. https://www.news24.com/news24/video/southafrica/news/documentary-slain-by-soldiers-the-collins-khosa-story-20210327. Accessed 8 Jan 2023
Schindler S (2017) Towards a paradigm of southern urbanism. City 21(1):47–64
Soja E (2009) The city and spatial justice. Justice spatiale/Spatial Justice 1(1):1–5

Soja E (2010) The city and spatial justice. In: Bret B, Gervais-Lambony P, Hancock C, Landy F (eds) Justices et injustices spatiales. Presses Univeritaires de Paris Quest, Paris, pp 55–74

Stivers C (2007) "So poor and so black": hurricane Katrina, public administration, and the issue of race. Public Adm Rev 67:48–56

Tulchinsky TH (2018) John snow, cholera, the broad street pump: waterborne diseases then and now. Case Stud Public Health 2018:77–99

UNAIDS (2020) Rights in a pandemic: lockdowns, rights and lessons from HIV in the early response to COVID-19. UNAIDS, Geneva

UNDP South Africa (2020) Impact of COVID-19 on informal and small businesses in South Africa. UNDP South Africa, Pretoria

United Nations (1945) Charter of the United Nations and Statute of the international court of justice. United Nations, San Francisco

Wanjiru MW, Matsubara K (2017) Stret toponymy and the decolonization of the urban landscape in postcolonial Nairobi. J Cult Geogr 34(1):1–23

Wekwete KH (1988) Development of urban planning in Zimbabwe: an overview. Cities 5(1):57–71

Yiftachel O (2009) Theoretical notes on 'gray cities': the coming of urban apartheid? Plan Theory 8(1):88–100

Young IM (1990) Justice and the politics of difference. Princeton University Press, Princeton and Oxford

Zenda C (2020) Army and police brutality keeps starving Zimbabweans indoors. https://www.fairplanet.org/story/army-and-police-brutality-keeps-starving-zimbabweans-indoors/. Accessed 9 Jan 2023

Chapter 3
Contextualizing Harare Urban Socioeconomic Profile and History of Pandemics in the City

Introduction

Across the world, cities represent a better life with opportunities for employment and access to healthcare and education; as a result, they attract people living in rural and peri-urban areas. The world's urban population grew rapidly from 751 million in 1950 to 4.46 billion in 2021 (United Nations 2018). It is also estimated that the population will rise to 6.68 billion by 2050, with an additional 2.22 billion people in urban areas and 90% of the increase expected to be in Asian and African countries (United Nations 2018). New megacities—cities that exceed a population of 10 million—such as Kinshasa[1] in Democratic Republic of Congo (DRC) and Lagos[2] in Nigeria are emerging in Africa (Hotez 2017). While growth is associated with rural–urban migration, the existing population also rose unexpectedly. Many scholars welcome such transformation as a positive development (Hayes 2014; Potts 2012; Turok 2014). Likewise, city planners and urban geographers have for decades argued that denser concertation of people promotes more efficient utilization of resources, therefore reducing the impact on the environment and promoting sustainable development (Berg 2016). Some studies link urbanization with socioeconomic and cultural productivity, translating into greater life expectancy, increased access to education facilities, and better living conditions (United Nations 2014). Without detailing the positive changes promoted by

[1] In 2017, the most recent population estimate for the city, Kinshasa had a population of 11,855,000. The population of metropolitan Kinshasa is projected to increase significantly to 35 million by 2050, 58 million by 2075, and 83 million by 2100, making it one of the largest metropolitan areas in the world (see Hoornweg and Pope 2017).

[2] Lagos is the most populous city in Africa with an estimated population of more than 24 million in 2022 and around 30 million for the Lagos metropolitan area (see United Nations World Population Prospects 2018).

J. I. Bhanye et al., *COVID-19 Lockdowns and the Urban Poor in Harare, Zimbabwe*, https://doi.org/10.1007/978-3-031-41669-9_3

increasing urbanization, cities are also associated with adverse effects on the urban population, especially in the Global South.

Urbanization can expose populations to risk of pathogens and the massive spread of emerging and reemerging infectious diseases. Globally, cities are the most hit by pandemics because they are characterized by high population density, which defines the socioeconomic space of the cities. Rapid population growth can also overwhelm cities' capacity to sustain healthy populations by providing clean water, sewage management, quality housing, unpolluted environments, and healthy diets. For example, when the Zika virus hit Brazil in 2015, a dense population was established as a conduit of the virus in Rio de Janeiro, making the pandemic a pernicious city problem (Berg 2016). Given that increased population in a city is associated with increased mobility, in western Africa and Asia, specifically in Hong Kong, the Ebola virus diseases (EVD) and severe acute respiratory syndrome (SARS), respectively, exponentially spread due to high travel of people within the cities and outside including regional and international (Berg 2016). Likewise, the COVID-19 pandemic spread was closely associated with the movement of people from place to place, locally, nationally, regionally, and internationally.

While the spread of pandemics, especially respiratory diseases, is often perceived as nondiscriminatory, more scholars are associating the spread of pandemics in cities with inequalities and urban poverty (Bhanye and Bhanye 2023; Chirisa et al. 2020). The impacts of pandemics are severe among the urban poor, who are defined by slums, poor sanitation, low-quality housing, and crowding, which enhance spreading of viruses and pathogens. Snyder et al. (2017) argued that the sporadic spread of the Zika virus was more visible in densely populated informal urban settlements of Latin America and the Caribbean. Similarly, mortality trends of influenza (Spanish flu) in Europe were high in poor, polluted urban parishes compared to the least polluted parishes; estimates indicated an increase of 1.6 deaths per 1000 population in poor urban parishes, which could have been reduced by 4000 if all communities had equal conditions. Recently, COVID-19 has clearly shown the connection of pandemics with mobility, as it spread within and between cities and city regions both globally and within countries (Harris et al. 2022). The pandemic also reflected the socio-spatial unevenness of urban areas since it followed historical trajectories that shaped urban design. This was more visible in the Global South.

Urban dwellers in southern Africa face severe risks because of structural inequalities associated with the origins of African cities. It is widely acknowledged that African cities are colonial scars that were developed to strengthen colonial dominance, thus less cognizant of the socioeconomic and cultural aspects of Black Africans during that time and postcolonially. In colonial times, White settlers often invoked seclusion policies to access cheap labor, including during pandemics. When the third bubonic plague pandemic hit South Africa in 1901 after spiking in Hong Kong in 1894, the colonial government amended the Public Health Act of 1897 to expel Black Africans from cities, particularly Cape Town, and relocate them to sewage farms—*the native location*—as a way of controlling the spread of the pandemic but retaining access to cheap African labor (Swanson 1977). While in postcolonial southern Africa, vulnerability to pandemics is a combination of multiple factors. Currently, this region faces a distinct set of challenges for pandemic control, including a lack of public infrastructure and access to adequate healthcare, housing, and sanitation.

In the last decade, the WHO recognized that socioeconomic and settlement unevenness in urban areas promotes the spread of pandemics including neglected transmission diseases and noncommunicable diseases (WHO 2016; Hotez 2017). Likewise, more recent discussions on African urbanization acknowledge the signing of the Sustainable Development Goals (United Nations 2015) and the New Urban Agenda (UN-Habitat 2017). Urbanization is also recognized as a cross-cutting development that impacts other Sustainable Development Goals, including health. Alongside this, the African Union's Agenda 2063: The Africa We Want endorses an urban vision that recognizes cities as critical zones where people have the right to access basic life necessities, specifically health. To emphasize this, the African Union set the Africa Health Strategy 2016–2030, which offers a cohesive and consolidative platform for the health sector, including cities. These efforts are based on the historical understanding that cities are the most hit by pandemics globally. However, little is known about who is the most vulnerable or what happens, for example, to the homeless, street vendors, and those in slums if the context of pandemics. Additionally, the structural aspects that determine urbanities' capacity to access health services during pandemics are not well-documented.

Against this background, this chapter contextualizes the history of pandemics with geospatial, socioeconomic, and institutional dynamics unfolding in African cities, using Harare as a case study. Harare has witnessed serial outbreaks of pandemics during the past two decades. The city's origins suitably fit several dynamics associated with histories of urbanization in the Global South and how such historical aspects still influence urban life (Zinyama and Tevera 2017). Additionally, in the past two decades, Harare witnessed multiple socioeconomic and geopolitical trajectories that are significant to consider when discussing urban life and pandemics in the Global South.

This chapter is organized into seven interrelated sections. The first contextualizes histories of urbanization and pandemics, using Harare as a case study. The second section explores the geohistorical background of Harare and how like cities in the Global South, and southern Africa in particular, is a product of colonialism. The third section discusses the colonial economy of the city shaped by dispossession, forced displacements, and forced labor driven by harsh colonial legal instruments. Part of the section reflects on the development of the city in postindependence Zimbabwe, with particular reference to how the new Black African-led government managed the city. The effect of how neoliberalism policies such as the Economic Structural Adjustment Programme of the International Monetary Fund and World Bank rendered many African urbanities poor and incapable of accessing social services including health is discussed. The section also considers environmental shocks of droughts and poor policies aimed at controlling cities' populations and structure. The fourth section discusses how a failing economy and geopolitical relations

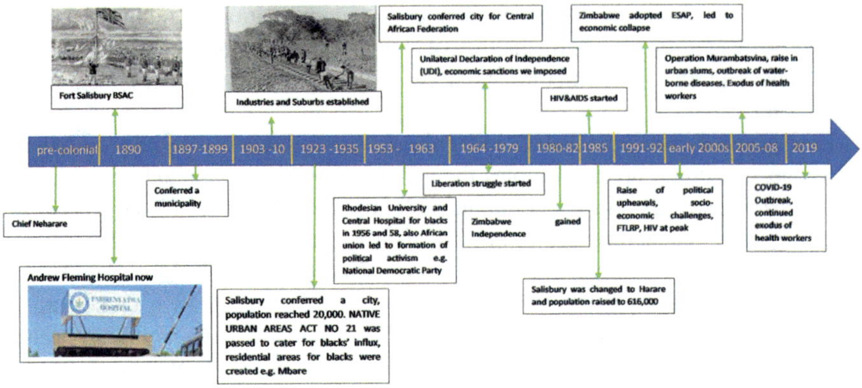

Fig. 3.1 Time series of major socioeconomic and political events in Zimbabwe. (Source: Authors)

shaped urban life in Harare and the country at large. Specifically, the section assessed how sanctions imposed on Zimbabwe by Western countries and the United States of America in 2000 promoted hardships in urban life, increased migration, and resulted in a crumbling social service delivery in the city. The fifth section discusses pandemics and epidemics that ravaged the city and how they were intertwined with socioeconomic, geopolitical, and governance issues unfolding in the country and the city of Harare. In the sixth section, we call for more robust research on how healthcare delivery can best be approached in cities of the Global South. While the section focuses on pandemics and infectious diseases, we call for a deeper understanding of how cities are designed and/or prepared to counter other diseases, including noncommunicable diseases. The seventh section offers concluding remarks on the troubled history and effect of pandemics in postcolonial Harare. Figure 3.1 shows a time series of major socioeconomic and political events in Zimbabwe from the precolonial period, colonial period, and postindependence, which impacted the history and current state of the city of Harare.

Background of Harare: Pre-independence Socioeconomic and Spatial Segregation

The context of this book is urban Harare. Harare is the capital and most populous city of Zimbabwe, situated in northeastern Zimbabwe in the Mashonaland region. Harare was founded in 1890 by the Pioneer Column, a small military force of the British South Africa Company, and named Fort Salisbury after the British Prime Minister Lord Salisbury. Salisbury was declared a municipality in 1897 (Zinyama et al. 1993). The British South Africa Company administrators demarcated and ran the city until Southern Rhodesia gained a responsible government in 1923. Salisbury was thereafter the seat of the Southern Rhodesian (later Rhodesian) government and, between 1953 and 1963, the capital of Southern Rhodesia and Nyasaland.[3] By 1935, the population had grown to nearly 20,000, and Salisbury was officially declared a city.

During colonial times, the city was segregated by class and race to promote white interests at the expense of the non-White population (Mlambo 2019). As a colonial city, there was limited space for Black Africans. Small towns were created for Black African workers employed as tea boys, factory workers, and agricultural laborers. These towns included Harari Township, now "Mbare," around the 1900s, and other locations like Highfields Township, established in 1937 (Gundani 2019). These townships were created following the passing of the Native Urban Locations Ordinance (No. 4 of 1907) by the British South Africa Company,[4] which prohibited "free residents" in Salisbury (Gundani 2019). The ordinance required "all Africans in Salisbury, except those already sleeping on employers" premises, to move to the "African location." It effectively made African residency in the town of Salisbury illegal (Yoshikuni 2007: 9).

The government of Southern Rhodesia introduced harsh laws to enforce racial segregation in Salisbury. In particular, the *Native Urban Areas Accommodation Act* of 1947 turned Salisbury into a "White city." Black African towns were created to accommodate employed men, thus commonly associated with bachelors' hostels and women were not allowed in the townships. According to the White colonizers, this helped control the male population in urban areas. Additionally, the ordinance was designed to ensure that all Africans lived in townships and not in the city, that African domestic servants should not be allowed to keep their families with them in the suburbs, that facilities be strictly segregated, and that "elite" African be corralled into small areas on the margin of townships (Chirisa 2010; Matamanda et al. 2022a, b, c). However, this segregating policy dehumanized an entire Black population and has become the source of all disparities, including those continued in post-colonial Salisbury.

At the time of Zimbabwe's independence in 1980, Salisbury's population was 616,000; it was renamed Harare on the second anniversary of Zimbabwean independence from British Colonial Rule. In 2012, the population of Harare was estimated to be 1.4 million, while the entire metropolitan area raised this figure to 2.1 million (Government of Zimbabwe 2012). The recent census report (2022) estimated the current population of Harare to be 2,427,209. Due to the increasing population and urbanization, Harare is a metropolitan province incorporating Chitungwiza and Epworth's municipalities. Regarding the residential settlements, Harare is divided into different classes of affluence—indications of perpetuating colonial

---

[3] The Federation of Rhodesia and Nyasaland, also known as the Central African Federation or CAF, was a colonial federation that consisted of three southern African territories: the self-governing British colony of Southern Rhodesia and the British protectorates of Northern Rhodesia and Nyasaland. It existed between 1953 and 1963.

[4] The British South Africa Company was the colonial administration in charge of Mashonaland since 1890 and Rhodesia between 1895 and 1923.

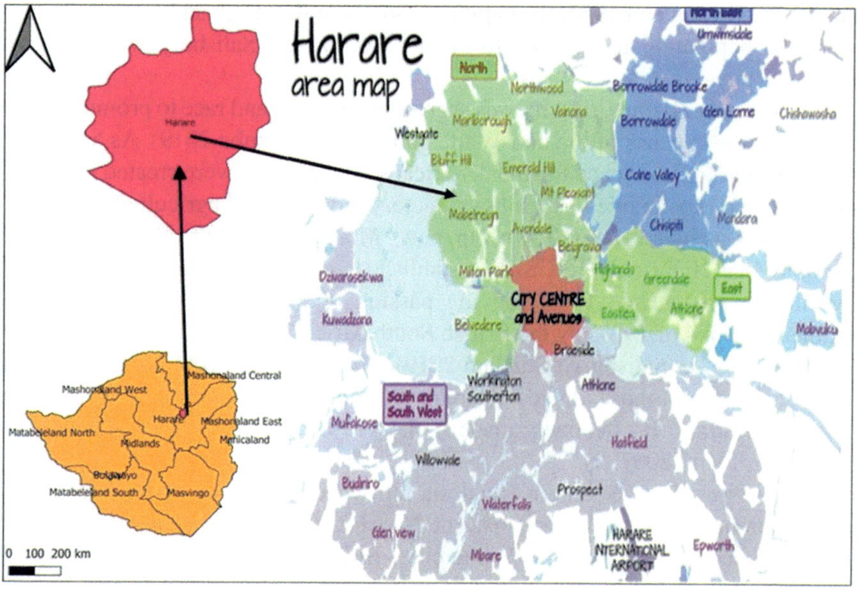

Fig. 3.2 Areas and suburbs of Harare, Zimbabwe. (Source: Authors)

spatial segregation of city dwellers. The northern and northeastern suburbs are home to the more affluent population. These suburbs are often called "dales" because of the common suffix -dale found in some suburbs such as Avondale, Greendale, and Borrowdale. The middle class reside in the suburbs like Belvedere, Bluffhill, Eastlea, and Milton Park. The south and southwestern suburbs, Dzivarasekwa, Kuwadzana, Mufakose, Budiriro, Glen View, Mbare, and Epworth, among others, house most of the urban poor who survive mainly through the informal sector. Figure 3.2 shows the areas and suburbs of Harare, the capital city of Zimbabwe.

Postindependence Urbanization Trends, Migration, and Population Displacements

Harare is the biggest primary city in Zimbabwe, followed by Bulawayo, Chitungwiza, Mutare, Gweru, Kwekwe, Kadoma, and Masvingo. Harare is the capital city where most of the country's political and economic processes are concentrated (Masimba 2021). The city is Zimbabwe's leading financial, commercial, and communications center and trade center for agricultural produce (cotton, tobacco, maize, and citrus fruits). Figure 3.3 is the map of Zimbabwe showing the location of the primary and secondary cities.

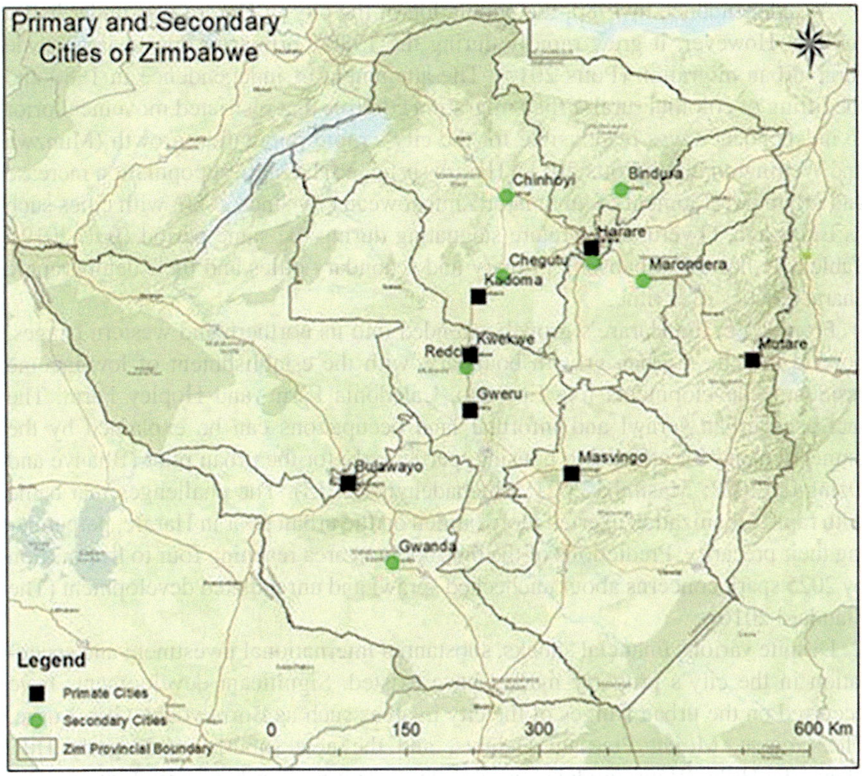

Fig. 3.3 Map of Zimbabwe showing the location of the primary and secondary cities. (Source: Authors)

The city is a major hub for the country's road, rail, and air transport networks and is strategically positioned for trade and tourism. After independence in 1980, Harare inherited a robust manufacturing sector anchored by mining and agricultural activities (Cheetham 2011). Estimates suggest that one in three Zimbabweans live in Harare, with the city's economy contributing 40% of the national gross domestic product (Government of Zimbabwe 2022). However, with nearly two decades of economic turmoil, the city's formal economic sector has shrunk significantly. For example, manufacturing, clothing, and textile industries have collapsed, with factories reduced to dilapidated shells (Kamete 2004). This resulted in the exponential growth of the informal sector. A staggering 90% of Zimbabwe's working population is estimated to be employed in this sector (Dube and Chirisa 2012). The informal sector covers trade in flea markets, home industries, social markets, guided car markets, and street vending. In Harare, traders often operate from insecure contested urban spaces like the downtown area, bus terminals, and along the streets, with inadequate infrastructure facilities.

At independence in 1980, the population of the city of Harare was under half a million. However, it grew rapidly during the 1980s, primarily due to large-scale rural–urban migration (Potts 2010). The attainment of independence in 1980 and the lifting of colonial rural–urban migration controls that restricted movement prior to independence was responsible for the city's rapid population growth (Munzwa and Wellington 2010; Potts 2011). Harare sustained the highest population increase and urban development of any major Zimbabwean city since 2000, with cities such as Bulawayo, Gweru, and Mutare stagnating during the same period (IOL 2019). Table 3.1 shows Zimbabwe's primary and secondary cities and their demographic characteristics over time.

From 2006, the Harare's growth extended into its northern and western fringes, beyond the city's urban growth boundary, with the establishment of low-income residential developments like Epworth, Caledonia Farm, and Hopley Farm. The increased urban sprawl and informal land occupations can be explained by the unmet demand for affordable housing, particularly for the urban poor (Bhanye and Dzingirai 2022; Masimba 2021; Muchadenyika 2015). The challenges that came with rapid urbanization exerted more burden on the urban poor in Harare, perpetuating their precarity. Predictions of the metropolitan area reaching four to five million by 2025 spark concerns about unchecked sprawl and unregulated development (The Standard 2016).

Despite various financial shocks, substantial international investment and speculation in the city's property market have existed. Significant developments have occurred on the urban fringes of the city in areas such as Borrowdale, Glen Lorne, The Grange, Mount Pleasant Heights, and the new suburbs of Hogerty Hill, Shawasha Hills, Bloomingdale, and Westlea, resulting in urban sprawl into nearby Mount Hampden, Ruwa, and Norton (McGregor 2014). In addition, areas such as Avondale, Eastlea, Belgravia, Newlands, and Milton Park have seen increased gentrification driven by speculation from expat Zimbabweans, which attracted other foreign buyers in high property prices and widespread rent increases (Writer 2014).

Aside from the rapid rural-to-urban migration contributing to the city's rapid urbanization, another migration challenge to Harare's economy is the persistent emigration of highly educated and skilled residents to countries like the United Kingdom, Australia, Canada, Ireland, and New Zealand, mainly due to the economic downturn and political unrest (Clayton and Manyena 2020). A study by McGregor and Primorac (2010) points out that "since 2000, there is a sense in which everyone (Zimbabweans in the diaspora and internally) experienced their moves as forced." The city's brain drain led to the decline of a local entrepreneurial class, an overstretched and declining middle class, and a dearth of employment opportunities outside the informal and public sectors (Chikanda 2011). Further, the city's working-class residents are also moving to neighboring countries like South Africa, Namibia, and Botswana (Campbell and Crush 2012).

Less well-off rural migrants readily replace highly educated and skilled Zimbabweans migrating from Harare. Estimates of the total number of Zimbabweans living abroad vary from one source to another. The United Nations estimates that 571,970 Zimbabweans were living in South Africa, the United

Table 3.1 Primary and secondary cities of Zimbabwe and their demographic characteristics over time

		Census population figures					Average annual growth		
		1982	1992	2002	2012	2017	1992–2002	2002–2012	2012–2017
Primary cities (Level I)	Harare	658,364	1,184,169	1436,000	1,468,767	1,723,394	+2.13%	+0.23%	+3.47%
	Bulawayo	495,317	620,936	676,650	655,675	642,000	+0.90%	–0.31%	–0.42%
	Mutare	75,358	131,808	170,466	188,243	205,539	+2.93%	+1.04%	+1.84%
	Kadoma	44,585	67,267	76,351	90,109	105,826	+1.35%	+1.80%	+3.49%
	Kwekwe	47,976	74,982	88,000	100,455	114,621	+1.74%	+1.42%	+2.82%
	Gweru	78,940	124,735	141,260	158,233	182,918	+1.32%	+1.20%	+3.12%
	Masvingo	30,523	51,746	58,000	88,554	104,426	+1.21%	+5.27%	+3.58%
Secondary cities (Level II)	Chinhoyi	24,309	42,946	56,794	79,368	95,101	+3.22%	+3.97%	+3.96%
	Bindura	17,966	21,167	38,001	44,033	53,889	+7.95%	+1.59%	+4.48%
	Chitungwiza	172,556	274,912	321,782	356,840	428,571	+1.70%	+1.09%	+4.02%
	Gwanda	4920	10,565	13,184	20,420	26,518	+2.48%	+5.49%	+5.97%
	Marondera	19,971	39,384	52,283	62,120	68,017	+3.28%	+1.88%	+1.90%
	Redcliff	22,109	27,994	32,346	35,924	41,373	+1.55%	+1.11%	+3.03%
	Chegutu	19,606	30,122	42,959	49,832	56,731	+4.26%	+1.60%	+2.77%
Total national population		7.6 million	10.4 million	11.95 million	13.1 million	14.2 million			
Total population of primary cities		1.4 million	2.25 million	2.64 million	2.75 million	3.08 million	+60.70%	+4.17%	+2.40%
Total population of secondary cities		281,437	447,090	557,349	648,537	770,200	+58.90%	+16.36%	+3.75%

Source: Matamanda et al. (2022b)

Kingdom, Malawi, Australia, and Botswana in 2013. Ratha et al. (2011) provided another estimate of Zimbabweans abroad. These authors estimated that Zimbabwean emigrants numbered 1,253,100, constituting 9.9% of the population. Another, more credible, estimate comes from Statistics South Africa (STATS SA). Drawing on the 2016 Community Survey, STATS SA estimates that there were 574,047 Zimbabweans in South Africa. These numbers exclude a believed-to-be-large proportion of irregular-status Zimbabweans in the country. The Government of Zimbabwe (2022) Zimbabwe National Statistics Agency Census Report estimated the number of Zimbabweans living in the Diaspora to be around 900,000, while those living in South Africa alone are about 700,000. This means four out of five Diasporan Zimbabweans live in South Africa. Figures circulating in the media purport that the number of Zimbabweans living outside Zimbabwe varies significantly from four to seven million. However, it is generally accepted at over five million people, some 30% of all Zimbabweans, have emigrated.

Population displacements are commonplace in Zimbabwe. Since its independence in 1980, Zimbabwe has had two waves of out-migration (Venables 2020). The first wave of displacement occurred immediately after independence when many people left to avoid the new government. A mass exodus began in 2000 due to the increasingly punitive actions of the liberator-turned-dictator, President Robert Mugabe (Venables 2020). In the past 19 years, over three million Zimbabweans have fled due to the persecution of some ethnic groups and individuals as well as the economic and political decline of the country. Some were victims of Murambatsvina (Shona word for "drive out trash or clear the filth"); others, both Black and White farm workers and owners, fled because of farm invasions. Operation Murambatsvina, also officially known as Operation Restore Order, was a large-scale government campaign to forcibly clean up a huge slum that posed a public health risk due to lack of sanitation and services (Tibaijuka 2005). In May 2005, the Zimbabwean government demolished shanties, illegal vending sites, and backyard cottages in Harare, Epworth, and other cities during Operation Murambatsvina. However, it was alleged that the true purpose of the campaign was to punish the urban poor for supporting the opposition party, the Movement for Democratic Change, and to reduce the likelihood of mass action against the government by driving people out of the cities (Tibaijuka 2005).

Operation Murambatsvina affected at least 700,000 people directly and 2.4 million indirectly, leading to out-migration, internal displacement, and homelessness (Tibaijuka 2005). In 2000, Zimbabwe planned a land redistribution act to collect White-owned commercial farms attained through colonization and return the land to the Black majority. On February 26–27, 2000, the pro-Mugabe Zimbabwe National Liberation War Veterans Association (ZNLWVA) organized people (including but not limited to war veterans; many of them were their children and grandchildren) to march on White-owned farmlands, initially with drums, song, and dance (Pabst 2001). This movement was officially termed the Fast-Track Land Reform Programme. The predominantly White farm owners were forced off their lands along with their workers, typically of regional descent. This was done violently and without compensation. In this first wave of farm invasions, about 110,000 km^2 of land were seized (Sachikonye 2003). Several million Black farm workers were excluded from the redistribution, leaving them without employment. In addition, there was the hatred caused by Gukurahundi,[5] a massacre committed by the Mugabe-led government, the "regime's first and still unpunished genocide."

Postindependence Political and Economic Collapse

Following the Lancaster House Agreement in December 1979, the transition to majority rule in early 1980, and the lifting of sanctions, Zimbabwe enjoyed a brisk economic recovery. Real growth for 1980–1981 exceeded 20%. While Harare initially boomed under a wave of optimism and investment that followed the country's independence in 1980, in no time, the city began to gradually experience the impacts of political and economic collapse. The political and economic collapse was triggered by various shocks including the disastrous droughts of 1992 and 1995, the Economic Structural Adjustment Programme payment of war veterans, and joining the war in the Democratic Republic of Congo (DRC) in 1998. The major blow to the economy was witnessed following the Fast Track Land Reform Programme, which destroyed viable businesses and farms. The punitive Operation Murambatsvina (Clear the filth) resulted in the mass exodus of the poor out of the city. When Zimbabwe started to experience an economic downturn, the government responded by enacting neoliberal reforms such as the Structural Adjustment Programme in the early 1990s, which led to a boom in banking, finance, and agriculture. However, because of the neoliberal reforms, domestic firms struggled to compete with imports leading to the collapse of several institutions, particularly the textile industry (Mbiba 2017). Further, this ill-conceived macro-socioeconomic policy decision by the postcolonial government led to significant job losses in manufacturing, thereby significantly increasing unemployment and income inequality between the elite and the urban poor.

Starting in 1982, Zimbabwe experienced recurrent episodes of disruptive weather patterns cumulatively causing increased poverty across rural and urban population divides. Table 3.2 shows droughts in Zimbabwe between 1950 and 2013.

The country's worst drought happened in 1992, considered the most destructive drought Zimbabwe faced in the twentieth century. During this period, water shortages forced the shutdown of many industries and schools. Poor harvests that year

[5] Gukurahundi was a genocide in Zimbabwe, which arose in 1982 until the Unity Accord in 1987. The term derived from Shona language is loosely translated to "the early rain which washes away the chaff before the spring rains." It was framed by the government as the subduing of an insurgency in Matabeleland to maintain law and order. Gukarahundi was ethnic cleansing by another name and was responsible for an estimated 20,000 deaths.

Table 3.2 Droughts in Zimbabwe between 1950 and 2019

Grade	Extreme drought	Severe drought	Mild drought
Drought (years)	1983, 1992	1968, 1967, 1982, 2004, 2016, 2019	1951, 1960, 1964, 1965, 1970, 1984, 1987, 1991, 1995, 2002, 2003, 2005, 2007, 2009, 2010
Total	2	6	16
Percentage	8.3%	25%	66.7%

Source: Adapted from Nangombe (2015)

caused a ripple effect with decreased food production,[6] compromising food supply across southern African countries like Mozambique, which relied on exports from Zimbabwe—the breadbasket of Africa. A series of droughts occurred in 1995, 2002, 2003, 2005, 2007, 2009, and 2010. In 2016, Zimbabwe declared a drought disaster as an estimated five million people faced food shortages. The drought resulted from the shifts in weather patterns, a direct result of El Niño and La Niña, which refer to the periodic changes in sea temperatures in the Pacific Ocean. In 2019, Zimbabwe also experienced another severe drought, which ignited the worst hunger crisis the country had faced in nearly a decade, with only 55% of normal rainfall (Adames 2020). According to the World Food Programme, the 2019/2020 drought entered an Integrated Food Security Phase 3 crisis, just two steps above famine.[7] The droughts in Zimbabwe had many implications for the urban poor in cities like Harare. Implications ranged from crop failure for those practicing urban agriculture (Matamanda et al. 2022c), shortage in food and water supplies, and decreased household incomes. This rendered much of the urban population food insecure and poverty-stricken (Manjengwa et al. 2016).

Political economists argue that 1999 marked the era in which Zimbabwe's economy began to crumble visibly. The crash of the stock market on November 14, 1997, triggered alarm and the day was termed "Black Friday" due to the collapse of the Z$ against the US$ and British Pound and the plunge in the value of share prices listed on the Zimbabwe Stock Exchange (ZSE) (see Carmody 1998; Coltart 2008; Kanyenze 2004; Richardson 2005). Civil society groups began to agitate for their rights as these had been eroded under the Economic Structural Adjustment Programme. In 1997 alone, 232 strikes were recorded, the largest number in any

[6] Maize production dropped to a mere 361,000 tons in 1992, a figure significantly lower than the previous years production. Wheat production was also severely affected by the drought, with the country producing only 20,000 tons of wheat in 1992, a significant decline from the previous year's production of 100,000 tons. The drought also had a severe impact on Zimbabwe's livestock sector, particularly in the southern regions of the country.

[7] The global standard for measuring food insecurity is the Integrated Food Security Phase Classification (IPC) made up of five phases. These are IPC Phases 1 (Minimal), 2 (Stressed), 3 (Crisis), 4 (Emergency), and 5 (Famine) (https://www.ipcinfo.org/).

year since independence (Kanyenze 2004). During the first half of 1997, the war veterans organized demonstrations that the government initially ignored. As the intensity of the strikes grew, the government was forced to pay the war veterans a once-off gratuity of Z$50,000 by December 31, 1997, and a monthly pension of Z$2000 beginning January 1998 (Kanyenze 2004).

The Fast Track Land Reform Programme in the 2000s exacerbated the economic downturn. The new farm occupants, mainly indigenous citizens and several prominent members of the ruling Zimbabwe African National Union – Patriotic Front (ZANU-PF) administration, were inexperienced. Some were uninterested in farming, failing to retain the labor-intensive, highly efficient farm management practiced by previous White landowners (Stiff 2000). The current lack of agricultural expertise triggered severe export losses, a loss of market confidence, and a massive drop in food production. From 1991 to 2000, the production levels of tobacco, cotton, and soya bean showed an upward trend, though, for some years, such as 1992 and 1995, production levels fell because of drought. From the year 2000, when the Fast Track Land Reform Programme was implemented, production levels for tobacco, cotton, and soya bean showed a downward trend till the year 2019, except for tobacco, which showed an upward trend from 2008 to 2019 (Runganga et al. 2022). Figure 3.4 illustrates the production levels of major cash and food security crops in Zimbabwe from 1991 to 2019.

Maize production also showed a downward trend since implementing the Fast Track Land Reform Programme. Figure 3.5 shows the production levels for maize in Zimbabwe between 1991 and 2019.

Zimbabwe also experienced a currency crisis. At independence in 1980, Zimbabwe had a fairly stable currency, annual inflation was 5.4%, and

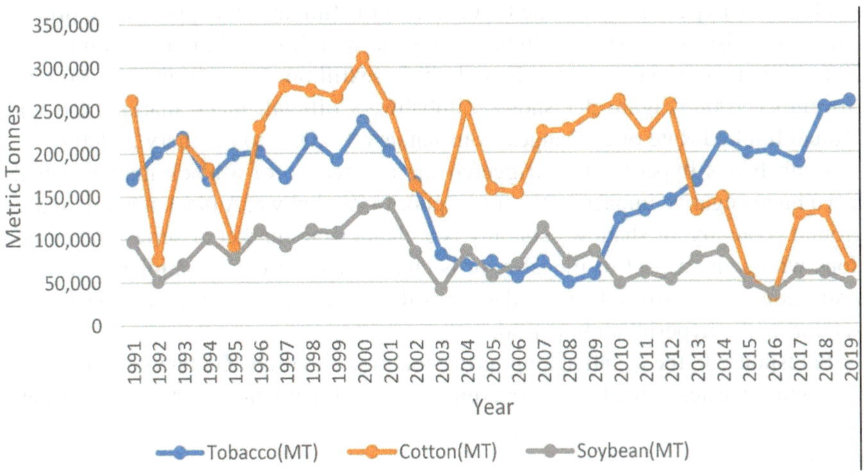

Fig. 3.4 Production levels of major cash and food security crops (1991–2019). (Source: Ministry of Land, Agriculture, Water, and Rural Resettlement Agricultural Statistical Bulletin, 2020)

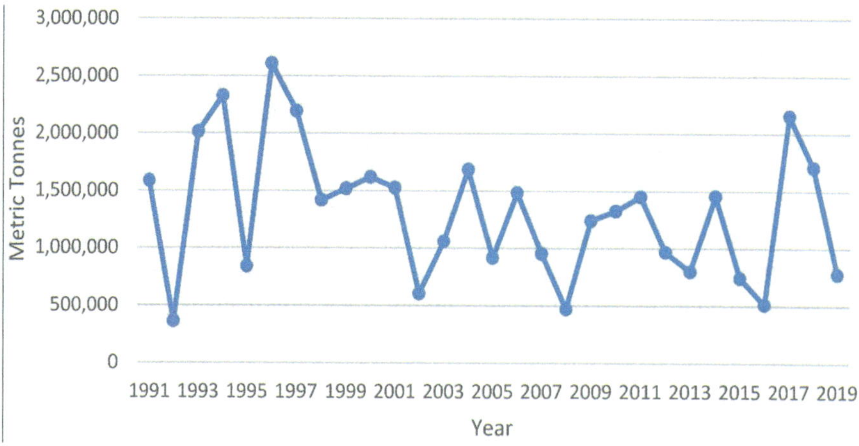

Fig. 3.5 Production levels for maize in Zimbabwe (1991–2019). (Source: Ministry of Land, Agriculture, Water, and Rural Resettlement Agricultural Statistical Bulletin, 2020)

month-to-month inflation was 0.5%. Currency of Z$2, Z$5, Z$10, and Z$20 denominations were released. Roughly 95% of transactions used the Zimbabwean dollar (Koech 2011). Zimbabwe began experiencing severe foreign exchange shortages, exacerbated by the difference between the official and black market rates in 2000. At the end of July 2008, the country's inflation rate hit a record high of 231 million % (MDTF 2008). Figure 3.6 shows bills during Zimbabwe's hyperinflation period ranging from Z$10 to Z$100 billion printed within 1 year. The magnitude of the currency scales signify the extent of the hyperinflation.

During the height of inflation from 2008 to 2009, it was difficult to measure Zimbabwe's hyperinflation.[8] The Government of Zimbabwe stopped filing official inflation statistics. However, Zimbabwe's peak month of inflation is estimated at 79.6 billion % month-on-month and 89.7 sextillion % year-on-year in mid-November 2008. Figure 3.7 shows Zimbabwe's official, black market, and Old Mutual Implied Rate of exchanges from January 1, 2001, to February 2, 2009.

Zimbabwe's hyperinflation triggered the emergence of a black market, which had a huge negative impact on the urban poor in Zimbabwe. A black market, underground economy, or shadow economy is a clandestine market or series of transactions with some aspect of illegality or is characterized by noncompliance with an institutional set of rules. Box 3.1 describes the situation in Zimbabwe's black market during the hyperinflation period.

In response to the negative long-term economic situation, Zimbabwe's three parliamentary parties agreed on a Government of National Unity 2009–2013. Despite

[8] Hyperinflation in Zimbabwe is an ongoing period of currency instability, which using Cagan's definition of hyperinflation, which began in February 2007.

Fig. 3.6 Bills during Zimbabwe's hyperinflation period. (Source: Authors)

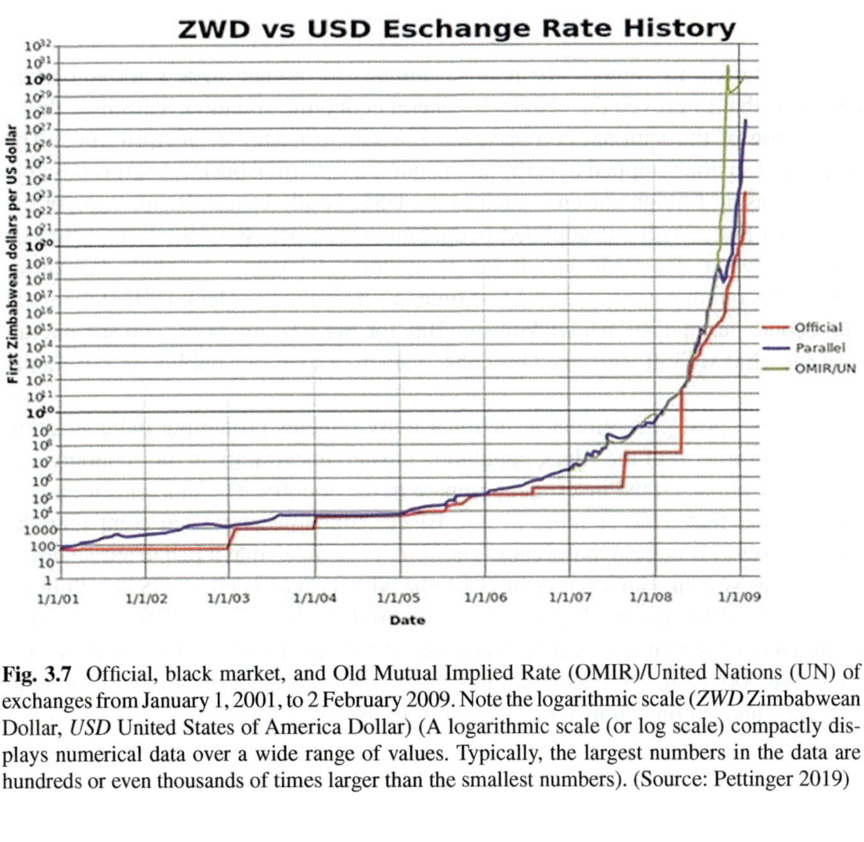

Fig. 3.7 Official, black market, and Old Mutual Implied Rate (OMIR)/United Nations (UN) of exchanges from January 1, 2001, to 2 February 2009. Note the logarithmic scale (*ZWD* Zimbabwean Dollar, *USD* United States of America Dollar) (A logarithmic scale (or log scale) compactly displays numerical data over a wide range of values. Typically, the largest numbers in the data are hundreds or even thousands of times larger than the smallest numbers). (Source: Pettinger 2019)

Box 3.1: Situation on the Zimbabwean Black Market
Prices in shops and restaurants were quoted in Zimbabwean dollars but adjusted several times a day. Any Zimbabwean dollars acquired need to be exchanged for foreign currency on the parallel market immediately, or the holder would suffer a significant loss of value. For example, minibus drivers were required by law only to accept payment from passengers in Zimbabwean dollars. However, at increasing rates throughout the day, the evening commute was the highest-priced ride of the day, with the next morning's price still higher. A driver might have to exchange money thrice daily, not in banks but in back-office rooms and parking lots.

Such business venues constituted a black market, an arena explicitly outside the law. Transactors could evade the price freezes and the mandate to use Zimbabwean dollars. The black market served the demand for daily goods such as soap and bread, as grocery stores operating within the law no longer sold items whose prices were strictly controlled or charged customers more if they were paying in Zimbabwean dollars. At one point, a loaf of bread was Z$550,000,000 in the regular market, when bread was even available; apart from a trip to another country, the black market was the only option for almost all goods at up to Z$10,000,000,000 for a loaf of bread.

Source: van Gelder (2008)

serious internal differences, the national unity government made some important decisions that improved the country's general economic situation, starting with the suspension of the national currency, the Zimbabwean Dollar, in April 2009. The strategic decisions stopped hyperinflation and made normal business forms possible again by using foreign currency such as the US$, South African Rand, Euro, or the Botswana Pula. With a disciplined budget and the expertise of the former finance minister Tendai Biti (Movement for Democratic Change), in 2009, Zimbabwe recorded economic growth for the first time in a decade (The Herald 2018). Political and economic analysts hailed Minister Biti for his sterling work, and the mantra "**we eat what we kill**" informed his conservatism in spending public funds from the treasury.

However, post-government of national unity policies promoting indigenization and empowerment of the Black population created uncertainty in the economy and negatively impacted the investment climate in the country. The indigenization policy required that 51% or more of non-Black Zimbabwean-owned companies be handed over to Black Zimbabweans. The country realized that it had to attract direct foreign investment. However, as of 2014, the country had a debt of US$7 billion, or over 200% of the country's gross domestic product. In August 2016, the government announced that it would be laying off 25,000 civil servants (8% of the country's 298,000 civil servants), cutting the number of embassies and diplomatic expenses and cutting ministerial expenses in an attempt to save US$4 billion in annual wages and secure help from the World Bank and the International Monetary Fund (Raath 2016). This move resonates with government policies in the mid-1990s that focused on retrenchment, which were introduced through the Economic Structural Adjustment Programme. In mid-July 2019, inflation had increased to 175% following the adoption of a new Zimbabwe dollar and banning the use of foreign currency, thereby sparking fresh concerns that the country was entering another period of hyperinflation (Samaita 2019). After the 2017 coup d'etat that ended the rule of Robert Mugabe in Zimbabwe, sentiments expressed in the inaugural speech by his successor Emmerson Mnangagwa suggested that he was going to adopt reforms to remove hindrances to economic progress, which had arisen during Mugabe's premiership.

On November 29, 2017 Mnangagwa said: "the liquidity challenges which have bedeviled the economy must be tackled head on, with real solutions being generated as a matter of urgency." He also announced a 3-month amnesty for the return of illegal externalized money and assets. Mnangagwa also promised to strengthen international engagement and promote democracy saying:

> Through the engagement and re-engagement policy, we are opening a new chapter in our relations with the world, underpinned by mutual respect, shared principles and common values. We look forward to playing a positive and constructive role as a free, democratic, transparent and responsible member of the family of nations (Mnangagwa 2018).

However, the new administration's ties to the old regime continued to maintain most of the old policies and practices. This resulted in failure to deliver on the promises of jobs, a growing economy, and a new and unfolding democracy for Zimbabwe.

Urban Politics and Distributive Injustices for the Urban Poor

The continued postcolonial political and economic collapse in Zimbabwe triggered a rise in urban politics and distributive injustices concerning urban space in Harare and other towns and cities. For nearly three decades, Harare's politics has remained hotly contested since the formation of the country's most prominent opposition party, the Movement for Democratic Change, now rebranded to Citizens Coalition for Change (Marumahoko et al. 2020; McGregor and Chatiza 2020). While social and physical infrastructure has been adversely affected in Harare, land access has been the most politicized (McGregor 2013; Moyo and Yeros 2007). Land and tribal issues have been at the heart of the country's populist politics and have dictated the land redistribution agenda (Chirisa et al. 2015; Muchadenyika 2015). The opposition dominates urban councils, while the ruling ZANU–PF's power is rooted in the countryside. Land allocations across the rural–urban divide have thus been politicized, with the opposition accused of favoring its urban constituency (Muchadenyika 2015). To counter this, the ruling party has established its loyalists within the city, allowing them to illegally occupy council or state land without the requisite town planning approvals and basic services.

The ruling ZANU–PF party has been accused of using national-level institutions, especially the Ministry of Local Government, to influence local authorities and even to invoke questionable ministerial powers to depose elected urban councils arbitrarily (McGregor 2013). The net effect of these local centered tensions of political configuration has been that national and local governments perpetually function at cross-purposes (Muchadenyika and Williams 2020). This has resulted in the continued deterioration of urban services, evident through frequent water cuts, sewage blockages, and dilapidated road infrastructure, especially in emerging settlements such as Hopley, Caledonia, and Hatcliff, where most of the poor reside (Bandauko et al. 2022; Matamanda 2020). The 2008 cholera outbreak in Harare, which resulted in nearly 5000 deaths nationwide, can be attributed to the city's urban crises and issues within social services and managed by urban councils (Venables 2020).

The intentional disruption of urban governance processes has, in turn, generated multiple informalities. Essential services such as rubbish collection and street repairs rapidly worsened and are now virtually nonexistent in poorer parts of the city. In May 2006, an editorial in the Zimbabwean newspaper, the Financial Gazette, described the city as a "sunshine city-turned-sewage farm" (FinGaz 2006). In 2009, Harare was voted to be the toughest city to live in, according to the Economist Intelligence Unit's Global Liveability Rankings.[9] The situation was unchanged in 2011, according to the same poll based on stability, healthcare, culture and environment, education, and infrastructure. The 2019 report by the Economist Intelligence Unit again dubbed Harare, Zimbabwe, one of the least livable cities in 2019. The report considered factors such as the economy, traffic congestion, infrastructure, conflict, and access to healthcare. Informal power structures that ZANU–PF either officially or unofficially control have emerged and set the scene for parallel urban service provision arrangements (Muchadenyika and Williams 2018; Venables 2020).

In addition, the self-provisioning of urban services by residents frustrated by the incapacitated urban councils has also led to the emergence of informal settlements (Muchadenyika 2014). For instance, old housing facilities built during precolonial times in suburbs like Mbare have seen the collapse of ageing water and sanitation systems (Matamanda et al. 2022a). In 2016, a citywide slum profiling and participatory mapping exercise produced an updated list of 63 informal settlements within the capital (Dialogue on Shelter Trust 2014). Yet, this was 10 years after the 2005 nationwide Operation Murambatsvina (Drive Out Filth),[10] which evicted and displaced 700,000 people in informal housing in urban centers (Tibaijuka 2005). More recently in 2019, Harare was ranked among the three provinces with the lowest

[9]The Global Livability Ranking is a yearly assessment published by the Economist Intelligence Unit (EIU), ranking 172 global cities (previously 140) for their urban quality of life based on assessments of stability, healthcare, culture and environment, education, and infrastructure.

[10]Operation Murambatsvina, also known as "Drive Out Filth," was a controversial government-led campaign in Zimbabwe that began in May 2005. The operation was officially aimed at restoring order and cleaning up illegal structures in urban areas, particularly in Harare, the capital city of Zimbabwe. However, the operation resulted in the destruction of informal housing and market structures, leaving an estimated 700,000 people homeless or without a means of livelihood. The operation was criticized by human rights organizations, including the United Nations, for its human rights violations and disregard for due process.

average household income out of the country's 10 provinces, with an average of US$55.31/month against an average total consumption poverty line of US$101.06 (Government of Zimbabwe 2019). These trends worsened due to the COVID-19 pandemic (World Bank 2020).

Harare's response to growing urban informality relied on evictions and demolitions as the default solution (Masimba 2021). This was demonstrated by the government-led campaign Operation Murambatsvina in 2005 and, more recently, COVID-19 lockdown-induced displacements (Bhanye 2023; Dialogue on Shelter 2020; Tibaijuka 2005). The same exclusionary practices have also befallen the informal economy in Harare (The Zimbabwe Mail 2020). Vending activities are viewed as a nuisance and criminalized rather than seen as productive processes contributing to the urban economy (Rogerson 2016; Toriro and Chirisa 2021). The City of Harare and the government used the cover of the coronavirus lockdown to tear down illegal market stalls in poor neighborhoods like Mbare. Citizens strongly criticized the move in the country where more than 80% of the working population have no formal jobs and eke a living from informal markets. City authorities defended the demolitions as necessary to restore order in the city and that informal traders would be relocated to new and better facilities. Defending the ongoing demolitions in Harare Local Government, Minister July Moyo said:

> Government is saying we cannot continue to just look at chaos in our cities if we want investment to come; if we want to reach vision 2030.

While the country's precolonial urban history accounts for segregationist planning responses, there has also been ill-informed political imperatives within postindependence city authorities, making them gravitate towards modernist planning ideals, which are often not aligned with prevailing socioeconomic circumstances (Chirisa and Dumba 2012; Kamete 2013; Zeiderman et al. 2017). Idealistic city visioning processes that set impractical minimum housing and infrastructure standards are classic examples. Box 3.2 describes the demolitions of "illegal" vending structures in Harare and Chitungwiza during the COVID-19 pandemic.

Housing demolitions have also increased in Harare in areas like Budiriro, Caledonia, Harare South and Kambuzuma, Zimre Park, and Chitungwiza (Bhanye

Box 3.2: Harare, Chitungwiza Kick-Start Demolitions During the COVID-19 Pandemic

Hundreds of informal traders in Harare and Chitungwiza were left without livelihoods after their councils joined forces with the police to demolish "illegal" vending structures across the capital and the dormitory town as per government directive. The demolitions were carried out despite residents' associations filing an urgent high court application seeking to bar government and local authorities from implementing the exercise.

In Chitungwiza's Zengeza 4, structures were demolished at the same time as those at the Mbudzi traffic circle along Harare–Masvingo Highway, with

(continued)

Box 3.2 (continued)
the council officials saying the program would spread across the city. Household furniture was destroyed at Gomba business center in the dormitory town before council workers drove to Chaminuka Drive and Katanga to demolish more structures using earthmoving machinery. Government, through provincial development coordinator Tafadzwa Muguti, last week threatened to demolish all illegal structures across the city, including shops and car sales. Information permanent secretary Ndavaningi Mangwana said: "A joint operation between ZRP [Zimbabwe Republic Police] and local authorities removing illegal informal activities operating within road servitudes has started today in Harare. Traffic interferes with illegal structures at Mbudzi roundabout and along Seke road near new Zengeza 3 and 5, which have been removed." National police spokesperson Assistant Commissioner Paul Nyathi said the police only enforced law and order during the demolitions.

Source: Kachiko, News Day (2021)

2023). In 2017, over 700 houses in Eyestone, Harare South, were demolished by the Harare City Council on allegations of being illegal erections, leaving many poor urbanites stranded and in the open. This happened although residents claimed to be affiliate members of the Harare South Housing Apex Cooperative Society, which was legally recognized and allowed to offer stands (News of the South 2017). In 2015, Harare City Council demolished 200 illegal structures along High Glen Kambuzuma Roads whose owners belonged to 32 bogus housing cooperatives. The houses were built on state land for a hospital and other amenities. Although notices of demolitions were sent 6 months earlier, settlers ignored the notices because cooperative representatives promised to liaise with the council. The settlers and their families were left homeless and did not know where to stay (The Herald 2015).

In another case, all houses and developments around the Robert Gabriel Mugabe International Airport area, including Arlington Estate and Retreat suburb, faced demolition threats. The developments were said to have violated the Civil Aviation Act and the Airports Company of Zimbabwe Act. However, authorities (the Government and Harare City Council) were cognizant of the erection of the illegal settlements and could have acted earlier to prevent development from taking place. These occurrences evoke memories of the infamous Murambatsvina operation where thousands of houses were demolished in 2005 (News of the South 2017). Local lobby groups and watchdog groups like the Harare Residents' Trust denounced the demolition of houses by Harare City Council, arguing that demolitions were not the solution and that humane responses were required (Mawire 2021).

Rising Urban Informality and Urbanization of Poverty

The impacts of increased urban politics and distributive injustices triggered a continued rise in urban informality, marginalization of the urban poor, and urbanization of poverty (Bandauko et al. 2021; Bhanye 2022, 2023). Harare witnessed rapid urbanization, triggering manifold socioeconomic and political challenges and perpetuating the marginalization of the urban poor (Bandauko et al. 2021; Venables 2020). The built-up areas in Harare increased from 279.5 km^2 in 1984 to 445 km^2 in 2018, with most of the land use and land cover changes towards the southwest where areas of high-density and often informal residential developments have emerged (Marondedze and Schütt 2019). There has been a rapid increase in informal settlements in Zimbabwe's urban settings, and in 2018 the World Bank estimated that 33.5% of the urban population resided in informal areas. Meanwhile, citywide slum settlement profiles conducted under the Harare "slum upgrading project" classified 63 neighborhoods as slums (City of Harare 2016; Dialogue on Shelter 2020). Shallow wells and pit latrines substitute for water and sanitation services, respectively.

Existing planned suburbs have not escaped infrastructure challenges, and old water and sanitation systems in places such as Mbare (the central market area) have collapsed due to overcrowding (Masimba 2021). The Mbare hostels, for example, which have an estimated population of 70,000 people, are overcrowded, severely dilapidated, and have been declared unfit for human habitation by human rights groups (Mazingaizo 2021). Overcrowded, dirty, and dilapidated, from the outside, these hostels look like hell holes, the only refuge for people living on the edge of society. Figure 3.8 shows dilapidated Mbare hostels.

Fig. 3.8 Dilapidated Mbare hostels. (Source: Authors)

Urban poverty has become entrenched, and an urban livelihoods assessment established that out of 10 provinces, Harare had the highest proportion of households that consumed poor diets. The COVID-19 pandemic compounded the urban socioeconomic challenges (Chirisa et al. 2020; Hunter et al. 2020). The pandemic overstretched Harare health facilities, and physical distancing also proved impractical in Harare, with residents queuing for water at boreholes, especially in low-income neighborhoods (Chirisa et al. 2020). Despite these deteriorating conditions, Harare continues with visionary processes, which imagine "a worldclass city by 2025" (City of Harare 2016, 2017).

Street vendors flood the streets, especially in Harare, but are expected to be formally registered and move to designated places. The influx of vendors in the CBD of Harare is not a standalone process; it is both a symptom and result of economic problems currently facing the nation. The formal economy alone cannot absorb both skilled and unskilled labor. An estimated 20,000 street vendors operate in Harare. The vendors sell everything from sweet potatoes to new and used footwear to pirated CDs. Goods splay across the sidewalks, making it a minefield for pedestrians in a hurry. Figure 3.9 shows street vendors on the streets of Harare.

The government wants vendors in Harare to pack up their wares and register and pay rent for one of 6000 places in designated zones, as it claims: "We cannot afford to have a nation of vendors." Figure 3.10 shows police arresting street vendors in Harare CBD.

But the reality, as observed by social and economic experts, is that hardships afflicting Zimbabwe have reduced the country to one of "the vendors," with everyone trying to sell something to survive (Toriro and Chirisa 2021; Mazhambe 2017).

Fig. 3.9 Street vending in Harare. (Source: Abraham Matamanda 2022)

Fig. 3.10 Police arresting street vendors in Harare CBD. (Source: Authors)

In most streets in urban areas, scores of young men and women roam around selling an assortment of products, including cell phone recharge cards, vegetables, clothes, traditional herbs, and skin-lightening creams (Mukwazhi 2015). For vendors, the streets are preferable as they earn more money there and "make ends meet in a tough economic environment"; as one vendor lamented, "Even if some of the vendors want to rent these spaces, rents are too high, they cannot afford, so better they sell in the streets."[11] Another vendor had this to say, "I have got two children to look after. I am paying school fees. I am getting the money from the street. They must have mercy for us."[12] "That is where we get food; that's where we get rent. That is where to get everything,"[13] said another concerned vendor in Harare CBD. Another vendor who said he had a university degree but was forced to sell on the streets because of lack of job opportunities in the city said, "I do not have any direction to go. Give us jobs, and we will work. Most of us are qualified. We have degrees and diplomas; we are drivers and mechanics, but there are no jobs."[14] While the government and the local authority want to banish vendors from the streets of Harare, the move can spark outrage among hordes of hawkers who have poured into Harare's streets to eke out an honest living (Mashaya 2017). Box 3.3 narrates the government's intended move to ban illegal vendors flooding the streets of Harare.

[11] Interview with street vendor in Harare CBD, July 13, 2022.

[12] Interview with street vendor in Harare CBD, July 13, 2022.

[13] Interview with street vendor in Harare CBD, July 13, 2022.

[14] Interview with street vendor in Harare CBD, July 13, 2022.

Box 3.3: Government Moves to Ban Street Vending

The government banning vendors from Harare's streets can spark outrage among hordes of hawkers who have poured into Harare's streets to earn an honest living. This comes as Mugabe's administration faces criticism for failing to create jobs for the burgeoning young population. Ever since the City of Harare launched a crackdown on vendors, they have taken to selling their wares well into the night to evade municipal police and also make a better killing through quick sales to a homeward-bound workforce.

Mugabe said he was told that some senior ZANU–PF leaders were protecting the vendors because they feared losing the forthcoming 2018 elections. "I was talking to (Home Affairs minister Ignatius) Chombo yesterday saying why do we allow our roads to be grocery shops? I said we must give them designated areas so that they will sell their wares outside the roads and leave the roads accessible."

He told me that the vendors want to sell their products everywhere. "I said that we do not want that indiscipline; down with that indiscipline, they must go to designated areas. We do not want the Nigerian style. Harare must be the smartest city because it is our capital city," Mugabe said. His sentiments come as there are fears the country could sink to the economic depths of 2007/2008 when shops were empty, and inflation hit world-record levels, resulting in untold pain and suffering for most citizens. Many people turned to vend after losing their jobs, with at least over 85% of Zimbabweans seeking employment since Mugabe won the 2013 elections.

Source: Mashaya (2017)

The government and local authorities vilify street vendors in Harare for bringing menace to urban navigation, fueling criminal activities, breeding diseases, and failing to comply with formal registration procedures and tax payments. Studies have, however, shown that vending has become an integral part of the world's urban economies, contributing to vibrant markets and providing an assortment of affordable, accessible goods and services to urban consumers (Mataire 2017). The question is thus: If researchers have proved that street vending is beneficial to the economic, social, and cultural being of cities, why do most African cities continue advocating that the trade be outlawed? (Mataire 2017).

Pandemics and Epidemics in Harare, Zimbabwe

The Centers for Disease Control and Prevention (2012) defines an epidemic as "the occurrence in a community or region of cases of an illness, specific health-related behaviour, or other health-related events clearly in excess of normal expectancy." A pandemic is "an epidemic occurring worldwide, or over a very wide area, crossing international boundaries and usually affecting a large number of people." Although recent pandemics and epidemics have been deadly, the mortality rates from these outbreaks are generally far lower than past health crises, due to advances in medicine and infrastructure. The global influenza pandemic of 1918—known to history as the "Spanish flu"—infected an estimated 500 million people and killed as many as 100 million (Nickol and Kindrachuk 2019).

In 1957, the "Asian flu" pandemic killed between 700,000 and 1.5 million people, resulting in a gross domestic product loss of 3% in the USA, UK, Canada, and Japan (Breitnauer 2020). The "Hong Kong influenza" pandemic of 1968 resulted in 1.2 million infections, with $23–26 billion direct and indirect costs in the USA. In 1981, the human immunodeficiency virus/acquired immunodeficiency syndrome (HIV/AIDS) pandemic recorded more than 70 million infections, 36.7 million deaths, and a 2–4% annual loss in gross domestic product growth in Africa (National Academies of Sciences, Engineering, and Medicine 2017). Since then, tens of millions of lives have been claimed by other pandemics and epidemics that also wreaked havoc throughout the globe. What is common about pandemics and epidemics is that they are concentrated in towns and cities where there is high population density. This section presents the major pandemics and epidemics (HIV/AIDS, cholera, and typhoid) that have hit the city of Harare over the years.

HIV/AIDS Pandemic

Zimbabwe is one of the countries hit hardest by HIV and AIDS in sub-Saharan Africa. The beginning of the HIV/AIDS epidemic in Zimbabwe dates back to the mid-1980s, when recorded cases increased by more than 60% (Chingwaru and Vidmar 2018). In the 1990s, the country had one of the largest recorded number of cases of HIV/AIDS in sub-Saharan Africa. By 2000, an estimated 25% of the population lived with the virus (Terceira et al. 2003). As of 2019, nearly 1.4 million people were living with HIV in Zimbabwe, of which 84,000 are children under 15 years (EGPAF 2021). Figure 3.11 shows new HIV/AIDS infections and deaths per year in Zimbabwe from 1990 to 2015.

Currently, initial case numbers during this period are considered to be vastly deflated. This is due to sociocultural barriers and the fact that individuals can be asymptomatic for up to two decades before they experience the symptoms, which necessitate diagnosis and treatment. Among adults aged 15–64 years, the prevalence of HIV varies geographically across Zimbabwe, ranging from 11.4% in Manicaland to 22.3% in Matabeleland South (Table 3.3).

In urban areas, some populations are more vulnerable to the impacts of HIV than others. Women living with HIV/AIDS undergo many obstacles due to sociocultural constraints, and gender norms prevent women from accessing healthcare services (Skovdal et al. 2011). Infants are at risk of contracting the virus through mother-to-child transmission when HIV-positive mothers breastfeed without antiretroviral therapy (McCarthy et al. 2017). Table 3.4 shows Zimbabwe's key HIV/AID statistics (2021).

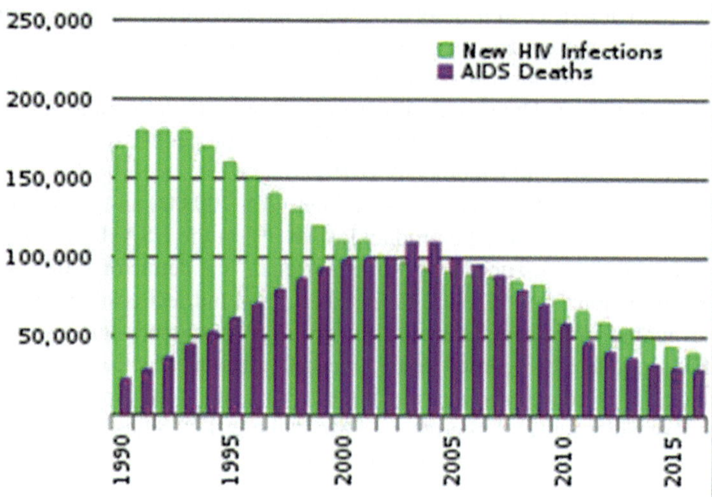

Fig. 3.11 New HIV/AIDS infections and deaths occurring each year in Zimbabwe from 1990 to 2015. (Source: UNAIDS)

Table 3.3 Prevalence of HIV varies geographically across Zimbabwe

Province	HIV prevalence (%)	95% confidence interval
Manicaland	11.4	9.9–12.9
Mashonaland Central	13.6	11.6–15.7
Mashonaland East	14.1	12.1–16.1
Mashonaland West	12.9	10.6–15.2
Matabeleland North	20.1	17.7–22.4
Matabeleland South	22.3	19.9–24.7
Midlands	14.1	12.0–16.1
Masvingo	14.9	12.6–17.2
Harare	14.2	12.6–15.8
Bulawayo	18.7	17.1–20.2

Source: Ministry of Health and Child Care (2016)

Table 3.4 Zimbabwe's key HIV/AID statistics (2021)

HIV/AIDS aspect	Statistics
People with HIV	1.3 million
Adult HIV prevalence	11.6%
New HIV infections	23,000
AIDS-related deaths	20,000
People on antiretroviral treatment	1.2 million

Source: Be in the Know (2021)

Despite its severity, HIV/AIDS transmission and prevalence in Zimbabwe have declined. Mortality rates in Zimbabwe attributed to the HIV/AIDS epidemic continue to decline along with diagnosed infections. As of 2018, UNAIDS reported that there was a 60% decrease in AIDS-related deaths since 2010, along with a 24,000-person decrease in new HIV infections (UNAIDS 2020). A review of the government's progress in implementing internationally recommended national treatment policies and implementation measures found that Zimbabwe had made substantial progress in its commitments (Tlhajoane et al. 2018). As of 2018, an estimated 1,086,674 people received antiretroviral treatment (CDCP 2019). However, access to treatment is still limited for certain groups and demographics, and data on specific populations' access to treatment are scarce due to legal barriers.

Preventing HIV/AIDS in Zimbabwe currently focuses on the prevention of mother-to-child transmission, voluntary medical male circumcision behavior change, comprehensive sexuality education (rolled out in primary and secondary schools), distribution of condoms (male and female condoms), and sexually transmitted infection management (including partner infections and PrEP, which is a daily administered pill, and a vaginal ring and injectable still to be introduced). The COVID-19 pandemic and associated lockdowns in Harare affected access to HIV prevention services, as neither family planning nor HIV testing was included as essential services (Frontline AIDS 2020). Restricted operating hours for public and private clinics and restricted public transport limited people's access to condoms, contraceptives, and sexually transmitted infection treatment. Despite HIV treatment being prioritized—the government agreed to a 6-month take-home supply for people living with HIV—there were stockouts of second-line antiretroviral medication. There were reports of high user fees, with people being charged US$3 for antiretroviral refills. The lockdown coincided with runaway inflation and health workers' strikes. The COVID-19 pandemic highlighted the need to find innovative approaches and differentiated service delivery models in cities like Harare, including community outreach for antiretroviral and PrEP refills, community distribution of safer sex commodities, and targeted HIV testing (Bhanye and Bhanye 2023; Frontline AIDS 2020).

Cholera Outbreaks

Cholera, an enteric infection caused by the bacterium *Vibrio cholerae,* causes diarrhea, leading to severe dehydration and death in people of all ages. Cholera is transmitted through ingesting food or water contaminated with *V. cholerae* and can lead to explosive, widespread epidemics (Chimusoro et al. 2018). Zimbabwe experienced cholera outbreaks dating as far back as 1971. More frequent outbreaks occurred in the late 1990s, with the largest recorded in 1999 when 4081 cases were reported in low-lying border areas covering six provinces (Chimusoro et al. 2018). Since 2000, cholera outbreaks are reported annually, with outbreaks occurring in 2008/2009 when 60 of the 62 districts in the country were affected (Mukandavire

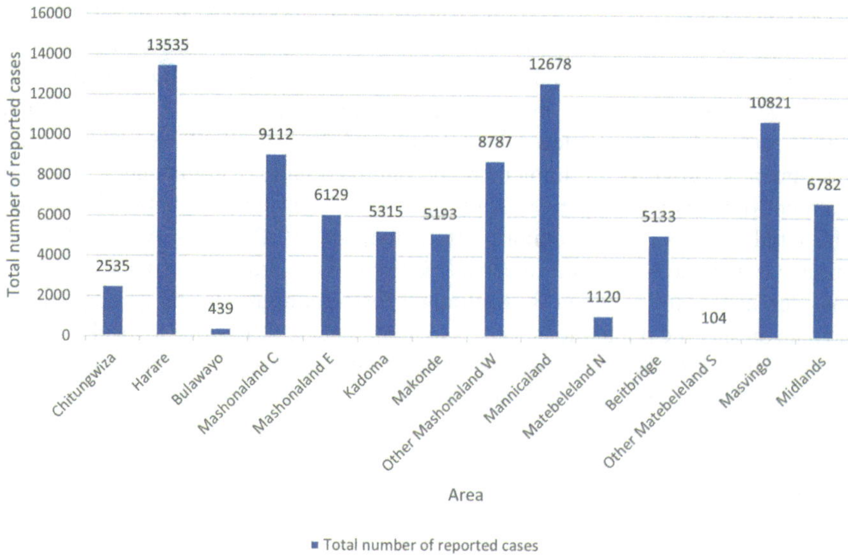

Fig. 3.12 The spread of cholera in Zimbabwe: The number of cases recorded in the most infected provinces of Zimbabwe as of March 4, 2009. (Source: World Health Organization 2019)

et al. 2011). The *V. cholerae* O1 El Tor outbreak of 2008–2009 was the largest recorded in Zimbabwe and one of the largest in Africa (Morof et al. 2013). The Zimbabwean Ministry of Health and Child Welfare reported an unprecedented 100,000 cases and nearly 5000 deaths from the catastrophic cholera outbreak (Chigudu 2019). Figure 3.12 shows the spread of cholera in Zimbabwe, the number of cases recorded in the most infected provinces of Zimbabwe as of March 4, 2009.

The health system in Zimbabwe, during the period 2008–2009, was far from resilient and unable to absorb shocks while maintaining normal functionality—characterized by a critical shortage of skilled as well as motivated health workers, critical shortages of essential medicines and supplies and medical technologies, dilapidated health infrastructure, unreliable health information systems and weak surveillance systems, poor service delivery, and poor health stewardship under inexperienced health leadership (Chimusoro et al. 2018). Since 2008, most of the cholera outbreaks in Zimbabwe were in poor urban settlements. The main drivers of cholera included the overloaded, dilapidated, and deteriorating water and sanitation infrastructure (Kone-Coulibaly et al. 2010). Figure 3.13 depicts poor sanitation conditions in Mbare, Harare, triggering cholera outbreaks.

Cholera outbreaks in Zimbabwe represent much more than public health crises but show the nadir of the country's deepening political and economic crisis. Hyperinflation, inadequate finances, and shortages of clinicians, medications, and supplies left Zimbabwe with an inadequate healthcare system, which was unable to respond adequately to a cholera outbreak. These risk factors contributed to the high death rate among the rural and urban poor. Cholera epidemics in cities like Harare have also been exacerbated by human rights abuses, including the politicization of

Fig. 3.13 Poor solid waste management and sanitary conditions in Mbare, Harare. (Source: Authors)

water, healthcare, aid, and information (Cuneo et al. 2017). Further, the cessation of water purification for the city and the redirection of raw human sewage into the city's main reservoir escalated the chances of cholera outbreaks. From September 1, 2018, to March 12, 2019, Zimbabwe experienced another large cholera outbreak, with 10,730 suspected cholera cases and 69 deaths (Mashe et al. 2020). More visible in Harare, the outbreak was linked to an inadequate supply of safe piped water and the subsequent use of unsafe water supplies (including boreholes and wells). Box 3.4 illustrates the experiences of the cholera outbreak, which took a toll on poor and overpopulated urban communities in Harare in 2018.

Box 3.4: Cholera Outbreak Takes a Toll on Poor and Overpopulated Urban Communities in Harare

Residents of high-density Harare suburbs Glen View and Glen Norah, at the center of the cholera epidemic, are in fear and uncertainty about whether they will survive the deadly scourge. Some schools have been closed and public events barred for fear that the disease will continue to spread; all these are parts of the measure to save people from the epidemic. The disease was first detected on September 1, and it has claimed many more than the conservative

(continued)

Box 3.4 (continued)
figure of 30 deaths announced by the government through the Ministry of Health and Child Care. "It is like we are on death row. You never know who the next victim will be, it is terrifying," said Shuvai Tembo of Glen View. She said her biggest fear was about her 10-month-old baby, noting it was now dangerous to leave her to crawl and play freely. "I now have to carry my baby Tinomuda most of the time for fear she might contract the disease while playing with others outside," she said. Tembo said that despite the difficult economic situation, she had to scrape together the money to buy sanitizers and other chemicals to protect her family.

Wadzanayi Sisimali, a vendor from Glen Norah, said life had become tough because she could no longer sell her wares. "I am at a loss, I cannot work, and I am also afraid of dying of cholera. For me, it is a double tragedy because I have lost my source of livelihood and live in fear of contracting cholera. I do not know what to do," said the single mother. Sisimali said that although she was aware of cholera's dangers, vendors needed to be provided with facilities to enable them to work for their families. A similar outbreak at the height of Zimbabwe's economic crisis killed more than 4000 people in 2008.

Investigations by Harare City Council have revealed that the major risk factors in the cholera outbreak are erratic water supply, burst sewer pipes, and the use of unprotected water sources. Further, due to shortages of foreign currency to buy chemicals, water purification has been a losing battle for most councils, resulting in residents drinking untreated water.

Source: Faji (2018)

Typhoid Outbreaks

Typhoid is caused by the bacterium *Salmonella enterica* serovar Typhi (*S.* Typhi), which is endemic in many parts of the world. Typhoid infection is transmitted via the fecal–oral route. It is more common among both urban slums and rural populations, which lack access to safe and adequate water, sanitation, and hygiene (Steele et al. 2016). Each year, typhoid fever causes an estimated 11–21 million cases and 128,000–161,000 deaths globally (WHO 2018). Due to run-down urban water and sanitation infrastructure, typhoid is endemic to Zimbabwe (Poncin et al. 2022). Zimbabwe has suffered regular outbreaks of typhoid, with most cases reported from the capital city, Harare (Polonsky et al. 2014), especially in the densely populated southwestern suburbs. Many cases have been reported since 2010, and seasonal outbreaks occur annually from October to March. Box 3.5 illustrates a case of typhoid outbreak in Zimbabwe in 2012.

Box 3.5: Zimbabwe: Five Deaths in Harare as Typhoid Continues to Spread

Zimbabwe's capital city was on high alert because of the spread of typhoid, with at least five related deaths registered in Harare since October 2012. The latest deaths have been recorded mainly in the Glen View suburb, where a fresh outbreak was reported in October, a year since the first outbreak in Harare in 2011. Harare City Health Deputy Director Dr. Prosper Chonzi said other new cases have also been reported in Mabvuku, Tafara, and other suburbs without access to clean water. "The fight against typhoid, which began in the city last year, is far from over as we continue to record fresh outbreaks," he said.

Typhoid cases have been reported in different parts of Zimbabwe since last year, with the worst affected areas being the densely populated suburbs around Harare's center, including Kuwadzana and Mufakose. More cases have been reported yearly in Bindura, Mashonaland Central, and Norton and Zvimba in Mashonaland West. Chitungwiza and Kadoma have also reported severe outbreaks, with the local authorities being blamed for failing to provide clean water. The latest outbreak in and around Harare brings the number of suspected cases registered nationwide to about 5000. The Health Ministry admitted it was not on top of the situation, with a critical lack of medicine and clean water hampering treatment and prevention efforts. Many local councils have also been unable to provide proper sanitation to their residents, blaming broken sewerage systems and water pipes for this failure. Key stakeholders contend that until there are lasting measures to combat water shortages, water provision, and sewage maintenance, diseases like typhoid will continue to be a threat.

Source: Munatsi (2012)

In October 2017, a typhoid outbreak started in Harare. As of June 28, 2018, 4330 cases were reported to the Ministry of Health and Child Care (Lightowler et al. 2022). The 0–14 year age group accounted for 47% of the cases. Between 2009 and 2017, 16,398 suspected and 619 confirmed cases (unspecified laboratory methods) of typhoid fever were notified (Mashe et al. 2019). The high-density southwestern suburbs of Harare, characterized by low socioeconomic status, intermittent water supply, frequent sewer line breaks, and low elevation, have historically been affected by typhoid fever, especially during the rainy season (Davis et al. 2018; Lightowler et al. 2022; Poncin et al. 2022). Figure 3.14 shows an example of poor sanitation conditions blamed for typhoid disease outbreaks in Harare.

Access to safe water and adequate sanitation and hygiene remain the mainstay of typhoid prevention and control. As an important complementary tool for endemic and epidemic disease control, the WHO recommends the programmatic use of typhoid vaccines, preferentially typhoid conjugate vaccine (TCV) (Lightowler et al.

Fig. 3.14 Poor sanitation conditions leading to typhoid disease outbreaks in Harare. (Source: Authors)

2022). In 2019 in Harare, the Government of Zimbabwe, through the Ministry of Health and Child Care and in collaboration with the WHO, Gavi, the Vaccine Alliance, and other partners, conducted a mass vaccination campaign of approximately 320,000 Typbar TCV doses targeting high-risk populations to control the outbreak and prevent possible peaks in the following seasons. The most affected suburbs (Mbare, Kuwadzana, and Glen View) were covered.

Zimbabwe is the first country in Africa to benefit from the mass typhoid vaccination, which has proven to be safe and effective. The Typbar TCV is new and, unlike routine typhoid immunization, is safe for children younger than 2 years (but older than 6 months) (WHO 2019). Table 3.5 shows the number of people vaccinated and administrative coverage by age group and suburb during the Typbar TCV campaign in Harare, Zimbabwe, in 2019.

Concluding Remarks: Troubled History and the Advent of Pandemics in Postcolonial Harare

Urbanization is widely acknowledged as a positive development that promotes human well-being. However, with the increasing (re) emergence of infectious diseases, cities are becoming uninhabitable. While there is a longstanding history of pandemics worldwide, the impacts thereof in cities is worrisome, especially in the Global South. In these cities, the urban poor are dying from these pandemics. This chapter, therefore, shows how pandemics are associated with socioeconomic inequalities in Harare. Before discussing how the urban poor are exposed to pandemics, this chapter presented the origins of Harare. Like other cities across the

Table 3.5 Number of people vaccinated and administrative coverage by age group and suburb during the typhoid conjugate vaccine campaign—Harare, Zimbabwe, 2019

Age group	Target population	Number vaccinated	Administrative coverage (%)
6 mos.–4 yrs.	114,388	82,768	72.4
5–15 yrs.	208,429	202,457	97.1
16–45 yrs.[a]	50,210	33,473	66.7
All	373,027	318,698	85.4
6 mos.–15 yrs. by suburb			
Budiriro	46,876	37,322	79.6
Dzivarasekwa	27,399	27,944	102.0
Glen Norah	28,251	28,436	100.7
Glen View	43,961	38,756	88.2
Hatcliffe	17,533	14,879	84.9
Hopley	43,924	34,182	77.8
Kuwadzana	62,268	56,453	90.7
Mbare	31,159	24,760	79.5
Mufakose	21,446	22,493	104.9
All suburbs	322,817	285,225	88.4

Source: Poncin et al. (2022)

[a]Vaccination among this age group took place in Mbare suburb only

Global South, the chapter reflected on Harare's inequalities that stem from colonial rule in Zimbabwe.

When the British settlers arrived in Zimbabwe in the late 1800s, they set up Salisbury, present-day Harare; the settlement was developed where colonialists could conduct administrative work to strengthen their territorial control. However, the settlement transformed into a town. Industrial areas, residential suburbs, social service facilities (health and education), and a commercial area (with banks) emerged. Through forced labor, Black Africans were assimilated into the city to provide cheap labor to the colonial masters. Men solely supplied labor. This social segregation shaped urban planning in the country and Harare specifically. Residential areas and social service facilities were divided into two distinct categories based on racial discrimination for Whites and Black Africans. The city was divided into two (uptown and downtown), where the former was explicitly reserved for Whites and the latter for Black Africans.

Additionally, Black Africans' residential suburbs were confined to the western part of the city, closer to industries; as a result, they often suffered from lung-related diseases triggered by exposure to industrial smoke. To Black Africans, accessing healthcare was a privilege. Two hospitals were established at that time: Parirenyatwa Hospital for Whites, Indians, and Coloured communities, while Gomo Hospital catered for the Black majority. Informed by these disparities, healthcare services and facilities in the two hospitals were sharply different. While no pandemic outbreak was recorded in Harare during that time, the way the city was structured significantly contributed to the spread and impact of pandemics 70 years later.

Harare was confronted by multiple socioeconomic and geopolitical challenges when the country came under Black majority rule, earned in the 1980s. With the excitement of gaining independence, the city witnessed increased population growth specifically due to rural–urban migration. A decade later, the new government adopted the Economic Structural Adjustment Programme, a set of infamous liberalization policies proposed by the Brenton Woods Institution. To align the economy with liberalism policies, the government withdrew support of social services (healthcare, education, social security, and provision of labor). On healthcare, the government reduced its support to public health facilities and encouraged private health service providers. This, however, burdened not only the urban population working in third-class jobs, but also urban women who were having their first experience of urban life. In the same decade, Harare and the country witnessed successive severe droughts. In 1992, for instance, the city witnessed food shortages and poor access to basic needs.

Compounded by these challenges, in the late 1990s, a new political party was formed by labor activists and considered by many as a savior of the urban poor in the city. Among its campaigning messages was to dismantle colonial exclusion, which were still omnipresent in postindependent Zimbabwe, including cities. However, the ZANU–PF-led government adopted the Fast Track Land Reform Programme to reward liberation fighters and masses with land to counter this. The results of unplanned land distribution were, however, unimagined; the country witnessed instant drops in the food supply since 70% of the country's economy hinged on agriculture. Politically, the eviction of White farmers was understood as a violation of human rights by several European countries, the United States of America, Canada, and Australia.

In return, economic sanctions were imposed on the country in early 2000.[15] However, this prompted significant economic collapse, which was more visible in the cities and Harare specifically. The decade was associated with increased closure and relocation of manufacturing companies and international banks. Thousands became unemployed, and there was a deterioration in the provision of social services, including education, water, sanitation and hygiene services, and healthcare. Harare, covered with indescribable filth, became the disaster's epicenter with poor sanitation services and congestion. Under these conditions, it witnessed infectious diseases, mainly waterborne such as cholera, diarrhea, and typhoid, due to poor sanitation. This overburdened the healthcare system already struggling to deal with HIV, sexually transmitted infections, and tuberculosis-related diseases that emerged in the late 1990s. These diseases were a cause of concern and health workers in many public hospitals fled the country due to poor working conditions, including the unavailability of protective clothing, also exposing them to the deadly diseases.

[15] The issue of sanctions is contested. One group (the Government of Zimbabwe and mainstream media) interpret them as economic sanctions. The other group (countries imposing sanctions and opposition politicians) call them targeted sanctions since they are directed to a selective group of politicians and companies that promote human rights abuses.

Despite these dynamics being well-captured in literature, we argued that not everyone died from these pandemics and was vulnerable to these diseases. In this chapter, we, therefore, illuminated that the urban poor residing in Harare were the most vulnerable to such pandemics. We reflected how ongoing inequalities engraved in colonial urban planning discourse informed how Harare urbanized, with elites occupying the formerly White northern suburbs commonly known as the "dales" where they accessed social services. At the same time, the poor continued to occupy the western suburbs that were congested with limited access to social services. We reflected on how life unfolded in these high-density suburbs by reviewing dynamics of Mbare—a colonial residential area—a hotspot of infectious pandemics. In short, the socioeconomic profile of Harare and the impacts of pandemics and epidemics demonstrated how urban crises triggered disparities and inequalities, penalized and perpetuated vulnerabilities of the urban poor, and threatened their rights to the city. While the recent outbreak of the COVID-19 pandemic posed new challenges to urban planning, we argued that more studies are imminent to trace how the impact of the pandemic was distributed on urban settlement patterns. This will improve knowledge on the spread of respiratory-related infections. Though the chapter focuses on infectious diseases, it is high time that urban planners consider healthcare and social services facilities to promote support to patients and elderly groups succumbing to communicable diseases.

Just like Harare, other African cities continue to face many crises such as political and socioeconomic failures, and inequalities to the impacts of climate change and pandemics (COVID-19 and HIV/AIDS). These crises have severe consequences for the urban poor, disproportionately affected by them. Crises and pandemics in African cities are products of a number of factors. Rapid urbanization, poor urban planning, and limited resources have created challenges such as inadequate housing, poor sanitation, and limited access to healthcare. These issues make African cities particularly vulnerable to the spread of diseases like cholera, typhoid, HIV/AIDS, and the recent COVID-19 pandemic. Additionally, many cities lack the resources to respond effectively to crises, including natural disasters and pandemics. Poor governance, corruption, and a lack of political will also contribute to the challenges facing African cities. Finally, climate change exacerbates these challenges, with rising temperatures, changing rainfall patterns, and more frequent extreme weather events, which put additional pressure on vulnerable urban communities.

African cities proved vulnerable to pandemics like COVID-19 due to a combination of factors. First, urbanization and rapid population growth lead to overcrowding, inadequate housing, and poor sanitation, thus creating ideal conditions for the spread of infectious diseases. Second, many African cities lack the necessary resources and infrastructure to respond effectively to pandemics, including limited healthcare facilities, inadequate testing and contact tracing capacity, and insufficient medical supplies. Third, poverty and inequality are widespread in many African cities, which limits access to basic services like clean water and sanitation and increases the risk of infection. Finally, weak governance and corruption hinder effective pandemic response, including delayed action, mismanagement of resources, and inadequate communication and coordination. Together, these factors made African cities particularly vulnerable to crises and pandemics like COVID-19.

This chapter contributes to ongoing debates by urban scholars who also demonstrated the volatile nature and vulnerability of African cities. The texts highlight the challenges facing African cities, including inadequate infrastructure, poor governance, and rapid urbanization. "Cities in contemporary Africa" edited by Martin Murray and Garth Myers focuses on the rapidly changing urban landscape of Africa. The book provides a comprehensive examination of the various forces shaping African cities, including globalization, urbanization, and political and economic changes. It explores the challenges facing African cities, such as poverty, inequality, and inadequate infrastructure, and discusses the ways in which African cities are adapting to these challenges (Murray and Myers 2007). "African Cities and the Development Conundrum" by Ammann and Förster (2018) examines the challenges facing African cities in terms of economic development, social inequality, and governance and explores the ways in which these challenges contribute to vulnerability in times of crisis. Gabriele Pasqui's recent book "Coping with the Pandemic in Fragile Cities" explores the effects of COVID-19 crisis on cities and urban areas and proposes approaches and solutions to invert the pandemic's negative impact (Pasqui 2022).

The book examines three aspects of the connection between pandemic and urban issues: the relevance of spatial and territorial variables in the explanation of pandemic dynamics and consequences in fragile cities; the assumption of radical uncertainty as the conceptual framework for a new approach to urban planning, in a rise of public investments; and the design of urban policies aimed at facing the material and symbolic effects of pandemics on the practices of use of spaces and places, in a context characterized by a plurality of populations and forms of life (Pasqui 2022). This chapter provided a background and for the book "COVID-19 Lockdowns and the Urban Poor in Harare, Zimbabwe," which focus on the impact of the pandemic on the urban poor and providing new perspectives on the morphing of a sustainable urban future in Zimbabwe and beyond.

References

Adames B (2020) How droughts in Zimbabwe affect the hunger crisis. The Borgen Project. https://borgenproject.org/droughts-in-zimbabwe/#:~:text=The%20country's%20worst%20drought%20happened,of%20many%20industries%20and%20schools. Accessed 21 May 2022

Ammann C, Förster T (2018) African cities and the development conundrum. Brill, Leiden

Bandauko E, Mutambisi T, Toriro P, Chirisa I (2021) Urban governance and the political economy of livelihoods and poverty in Harare, Zimbabwe. In: Wastl-Walter D, Kibicho JK, Owiso M (eds) Urban geography in postcolonial Zimbabwe: paradigms and perspectives for sustainable urban planning and governance. Springer, Cham, pp 77–92

Bandauko E, Kutor SK, Arku RN (2022) Trapped or not trapped? An empirical investigation into the lived experiences of the urban poor in Harare's selected informal settlements. Afr Geogr Rev:1–20. https://doi.org/10.1080/19376812.2022.2077781

Be in the Know (2021) At a glance: HIV in Zimbabwe. https://www.beintheknow.org/understanding-hiv-epidemic/data/glance-hiv-zimbabwe. Accessed 30 Jan 2022

Berg N (2016) As cities grow more crammed and connected, how will we discourage the spread of disease? Ensia. https://ensia.com/features/as-cities-grow-diseases-spread-faster-and-urban-design-key/. Accessed 12 Jan 2023

Bhanye J (2022) "Lydiate is now our home of a sort": perceptions of place amongst ageing first-generation Malawian migrants in Zimbabwe. Anthropol South Afr 45(3):180–194

Bhanye J (2023) 'Emerging forms of spatialised and socialised authority' among 'tenure-insecure peri-urbanites' in African peri-urban spaces: a review study. Pan-Afr Conversations 1(2):26–63

Bhanye J, Bhanye A (2023) A blessing or a curse?: The role of social media during the COVID-19 pandemic in Africa. In: The Palgrave handbook of global social problems. Palgrave Macmillan, Cham. https://doi.org/10.1007/978-3-030-68127-2_367-1

Bhanye J, Dzingirai V (2022) Structures and networks of accessing and securing land among peri-urban squatters: The case of Malawian migrants at Lydiate informal settlement in Zimbabwe. Afr Identities 20(3):237–256

Breitnauer J (2020) The Spanish Flu epidemic and its influence on history. Pen and Sword, Yorkshire

Campbell E, Crush J (2012) Unfriendly neighbours: contemporary migration from Zimbabwe to Botswana, Migration policy series no. 61. SAMP, Cape Town

Carmody P (1998) Neoclassical practice and the collapse of industry in Zimbabwe: the cases of textiles, clothing, and footwear. Econ Geogr 74(4):319–343. https://doi.org/10.2307/144468

Centers for Disease Control and Prevention (2012) Principles of epidemiology, 3rd edn. Centers for Disease Control and Prevention, Atlanta

Centers for Disease Control and Prevention (2019) Zimbabwe country profile. https://www.cdc.gov/globalhivtb/where-we-work/zimbabwe/zimbabwe.html. Accessed 12 Jan 2023

Cheetham RJ (2011) Infrastructure and growth in Zimbabwe: an action plan for sustained strong economic growth. African Development Bank Group, Tunis

Chigudu S (2019) The politics of cholera, crisis and citizenship in urban Zimbabwe: 'people were dying like flies'. Afr Aff 118(472):413–434. https://doi.org/10.1093/afraf/ady057

Chikanda A (2011) The engagement of the Zimbabwean medical diaspora, IDASA, migration policy series no. 55. SAMP, Cape Town. https://www.queensu.ca/samp/sampresources/samppublications/#policyseries. Accessed 13 Sept 2022

Chimusoro A, Maphosa S, Manangazira P, Phiri I, Nhende T, Danda S, Nabyonga-Orem J (2018) Responding to cholera outbreaks in Zimbabwe: building resilience over time. Curr Issues Glob Health 1(1):1–8. https://doi.org/10.5334/cigh.2

Chingwaru W, Vidmar J (2018) Culture, myths and panic: three decades and beyond with an HIV/AIDS epidemic in Zimbabwe. Glob Public Health 13(2):249–264. https://doi.org/10.1080/17441692.2017.1332718

Chirisa I (2010) Inner-city revitalization and the poor in Harare: experiences, instruments and lessons from Mbare. In: Urbanising Africa: the city centre revisited. Institute for Housing and Urban Development Studies, Rotterdam, p 58

Chirisa I, Dumba S (2012) Spatial planning, legislation and the historical and contemporary challenges in Zimbabwe: a conjectural approach. J Afr Stud Dev 4(1):1–13

Chirisa I, Bandauko E, Mutsindikwa NT (2015) Distributive politics at play in Harare, Zimbabwe: case for housing cooperatives. Bandung J Glob South 2(1):1–13

Chirisa I, Mutambisi T, Chivenge M, Mabaso E, Matamanda AR, Ncube R (2020) The urban penalty of COVID-19 lockdowns across the globe: manifestations and lessons for Anglophone sub-Saharan Africa. GeoJournal 87:1–14

City of Harare (2016) State of the city address – 1 March 2016 by his Worship Mayor Benard Manyenyeni

City of Harare (2017) City of Harare results based strategic plan (2017–2020)

Clayton J, Manyena B (2020) Diasporic reorientations: emotional geographies of the Zimbabwean diaspora in a post Mugabe era. J Ethn Migr Stud 48:1–18

Coltart D (2008) A decade of suffering in Zimbabwe. Development Policy Analysis, CATO Institute, Washington, DC

Cuneo CN, Sollom R, Beyrer C (2017) The cholera epidemic in Zimbabwe, 2008–2009: a review and critique of the evidence. Health Hum Rights 19(2):249

Davis WW, Chonzi P, Masunda KP, Shields LM, Mukeredzi I, Manangazira P, Appiah GD (2018) Notes from the field: typhoid fever outbreak—Harare, Zimbabwe, October 2016–March 2017. Morb Mortal Wkly Rep 67(11):342–343

Dialogue on Shelter (2020) Policy advice to respond to COVID-19 in urban informal settlements in Zimbabwe

Dialogue on Shelter Trust, Zimbabwe Homeless People's Federation and City of Harare (2014) Harare slum profile report, Harare Slum Upgrading Project

Dube D, Chirisa I (2012) The informal city: Assessing its scope, variants and direction in Harare, Zimbabwe. Global Advanced Research Journal of Geography and Regional Planning 1(1): 016–025.

EGPAF (2021) Elizabeth Glaser Pediatric AIDS Foundation (EGPAF) in Zimbabwe. https://www.pedaids.org/country/zimbabwe/. Accessed 13 Sept 2022

Faji T (2018) Cholera outbreak takes toll on poor and overpopulated urban communities in Harare. https://www.jozigist.co.za/cholera-outbreak-takes-toll-on-poor-and-overpopulated-urban-communities-in-harare/. Accessed 21 May 2022

FinGaz (2006) The Zimbabwe situation. https://www.zimbabwesituation.com/old/may19_2006.html#Z13. Accessed 20 Dec 2022

Frontline AIDS (2020) Zimbabwe HIV prevention shadow report 2020. https://hivpreventioncoalition.unaids.org/wp-content/uploads/2022/02/FrontlineAIDS_-Zimbabwe_FINAL_2.pdf. Accessed 13 Sept 2022

Government of Zimbabwe (2012) Zimbabwe census 2012 national report. Zimbabwe National Statistics Agency, Harare

Government of Zimbabwe (2019) Zimbabwe Vulnerability Assessment Committee (ZimVAC) 2019 urban livelihoods assessment

Government of Zimbabwe (2022) Zimbabwe National Statistics Agency Census Report. Zimbabwe National Statistics Agency, Harare

Gundani PH (2019) Church mothers of Mbare township: in memory of Mrs Elizabeth Maria Ayema (Mai Musodzi) and Sr Barbara Tredgold. HTS Teol Stud/Theol Stud 75(1):1–7

Harris P, Harris-Roxas B, Prior J, Morrison N, McIntyre E, Frawley J, Kwan B (2022) Respiratory pandemics, urban planning and design: a multidisciplinary rapid review of the literature. Cities 127:103767

Hayes M (2014) 'We gained a lot over what we would have had': The geographic arbitrage of North American lifestyle migrants to Cuenca, Ecuador. Journal of Ethnic and migration studies. 40(12):1953–1971.

Hoornweg D, Pope K (2017) Population predictions for the world's largest cities in the 21st century. Environ Urban 29(1):195–216

Hotez PJ (2017) Global urbanisation and the neglected tropical diseases. PLoS Negl Trop Dis 11(2):e0005308. https://doi.org/10.1371/journal.pntd.0005308

Hunter E, Price DA, Murphy E, van der Loeff IS, Baker KF, Lendrem D, Lendrem C, Schmid ML, Pareja-Cebrian L, Welch A, Payne BA (2020) First experience of COVID-19 screening of health-care workers in England. The Lancet. 395(10234):e77–8.

IOL (2019) Zimbabwe property market is characterised by a high demand and low supply. https://www.iol.co.za/pretoria-news/business-report/zimbabwe-property-market-characterised-by-a-high-demand-and-low-supply-20790629. Accessed 13 Sept 2022

Kamete AY (2004) Home industries and the formal city in Harare, Zimbabwe. Reconsidering informality: Perspectives from urban Africa. 1:120–38.

Kamete AY (2013) On handling urban informality in southern Africa. Geogr Ann Ser B Hum Geogr 95(1):17–31

Kanyenze G (2004) Economic Structural Adjustment Programme (ESAP): precursor to the fast track resettlement? Medicine Masiiwa, Friedrich Ebert Stiftung and Institute of Development Studies, University of Zimbabwe, Harare

Koech J (2011) Hyperinflation in Zimbabwe, Annual report. Globalization and Monetary Policy Institute, Dallas, pp 2–12

Kone-Coulibaly A, Tshimanga M, Shambira G, Gombe NT, Chadambuka A, Chonzi P, Mungofa S (2010) Risk factors associated with cholera in Harare City, Zimbabwe, 2008. East Afr J Public Health 7(4):311–317

Lightowler MS, Manangazira P, Nackers F, Van Herp M, Phiri I, Kuwenyi K, Luquero FJ (2022) Effectiveness of Typhoid Conjugate Vaccine in Zimbabwe used in response to an outbreak among children and young adults: a matched case control study. Vaccine 40:4199

Manjengwa J, Matema C, Tirivanhu D (2016) Understanding urban poverty in two high-density suburbs of Harare, Zimbabwe. Dev Sout Afr 33(1):23–38

Marondedze AK, Schütt B (2019) Dynamics of land use and land cover changes in Harare, Zimbabwe: a case study on the linkage between drivers and the axis of urban expansion. Land 8(10):1–20

Marumahoko S, Afolabi OS, Sadie Y, Nhede NT (2020) Governance and urban service delivery in Zimbabwe. Strategic Rev South Afr 42(1):41–68

Mashaya B (2017) Govt moves to ban street vending. Nehanda Radio. https://nehandaradio.com/2017/10/08/govt-moves-ban-street-vending/. Accessed 21 May 2022

Mashe T, Gudza-Mugabe M, Tarupiwa A, Munemo E, Mtapuri-Zinyowera S, Smouse SL, Mbanga J (2019) Laboratory characterisation of Salmonella enterica serotype Typhi isolates from Zimbabwe, 2009–2017. BMC Infect Dis 19(1):1–9

Mashe T, Domman D, Tarupiwa A, Manangazira P, Phiri I, Masunda K, Weill FX (2020) Highly resistant cholera outbreak strain in Zimbabwe. N Engl J Med 383(7):687–689

Masimba G (2021) Harare: city scoping study. African Cities Research Consortium. Retrieved January 14, 2023, from https://www.african-cities.org/wp-content/uploads/2021/12/ACRC_Harare_City-Scoping-Study.pdf. Accessed 21 May 2022

Mataire LR (2017) Is outlawing street vending the way to go. The Herald. https://www.herald.co.zw/is-outlawing-street-vending-the-way-to-go/. Accessed 21 May 2022

Matamanda AR (2020) Battling the informal settlement challenge through sustainable city framework: experiences and lessons from Harare, Zimbabwe. Dev South Afr 37(2):217–231

Matamanda AR (2022) Genesis of Hopley Farm informal settlement in independent Harare, Zimbabwe. African Geographical Review. 31:1–4.

Matamanda AR, Chirisa I, Rammile S, Marais M (2022a) Housing and technology: special focus on Zimbabwe, vol 37. Springer Nature, Cham

Matamanda AR, Mafuku SH, Bhanye JI (2022b) The potential of chinhoyi as a fast-growing secondary city in addressing urban challenges in Zimbabwe. J Asian Afr Stud. https://doi.org/10.1177/002190962211376

Matamanda AR, Mandebvu-Chaora C, Rammile S (2022c) The interplay between urban agriculture and spatial (in) justice: case study analysis of Harare, Zimbabwe. Land Use Policy 115:106029

Mawire W (2021) Housing demolitions by Harare City Council not the solution: says watchdog group. Pan African Visions https://panafricanvisions.com/2021/01/housing-demolitions-by-harare-city-council-not-the-solution-says-watchdog-group/. Accessed 21 May 2022

Mazhambe A (2017) Assessment of the contribution of street vending to the Zimbabwe economy. A case of street vendors in Harare CBD. IOSR J Bus Manage 19(9):91–100

Mazingaizo S (2021) Nearly 60 hostels in Zimbabwe's oldest and poorest township disconnected over unpaid bills. TimesLIVE, September 23. https://www.timeslive.co.za/news/africa/2021-09-23-nearly-60-hostels-in-zimbabwes-oldest-and-poorest-township-disconnected-over-unpaid-bills/. Accessed 21 May 2022

Mbiba B (2017) Harare: from a European settler-colonial 'sunshine city' to a 'zhing-zhong' African city. Int Dev Plan Rev 39(4):375–399

McCarthy E, Joseph J, Foster G, Mangwiro AZ, Mwapasa V, Oyeledun B, Essajee S (2017) Modeling the impact of retention interventions on mother-to-child transmission of HIV: results from INSPIRE studies in Malawi, Nigeria, and Zimbabwe. J Acquir Immune Defic Syndr (1999) 75(2):S233

McGregor J (2013) Surveillance and the city: patronage, power-sharing and the politics of urban control in Zimbabwe. J South Afr Stud 39(4):783–805

McGregor J (2014) Sentimentality or speculation? Diaspora investment, crisis economies and urban transformation. Geoforum. 1:(56):172–181.

McGregor J, Primorac R (eds) (2010) Zimbabwe's new diaspora: displacement and the cultural politics of survival, vol 31. Berghahn Books, New York

McGregor J, Chatiza K (2020) Partisan citizenship and its discontents: Precarious possession and political agency on Harare City's expanding margins. Citizenship Studies. 24(1):17–39.

Ministry of Health and Child Care (2016) Zimbabwe population-based HIV impact assessment ZIMPHIA 2015–2016. https://phia.icap.columbia.edu/wp-content/uploads/2016/11/ZIMBABWE-Factsheet.FIN_pdf. Accessed 21 May 2022

Ministry of Land, Agriculture, Water, and Rural Resettlement Agricultural Statistical Bulletin (2020) Agriculture, Land Reform, and Rural Development. https://www.dalrrd.gov.za/index.php/publications/45-statistics-and-economicanalysis/66-statistical-information. Accessed 3 Jan 2023

Mlambo AS (2019) Racism in colonial Zimbabwe. In: The Palgrave handbook of ethnicity. Palgrave Macmillan, London, pp 1–17

Mnangagwa E (2018) Inauguration speech by the incoming President of the Republic of Zimbabwe, Comrade Emmerson Dambudzo Mnangagwa National Sports Stadium, 26 August. https://irp-cdn.multiscreensite.com/ff7b7050/files/uploaded/HE%20INAUGURATION%20SPEECH.pdf. Accessed 28 May 2022

Morof D, Cookson ST, Laver S, Chirundu D, Desai S, Mathenge P, Handzel T (2013) Community mortality from cholera: urban and rural districts in Zimbabwe. Am J Trop Med Hyg 88(4):645

Moyo S, Yeros P (2007) The radicalised state: Zimbabwe's interrupted revolution. Rev Afr Polit Econ 34(111):103–121

Muchadenyika D (2014) Contestation, confusion and change: urban governance and service delivery in Zimbabwe (2000–2012) (Doctoral dissertation), University of the Western Cape

Muchadenyika D (2015) Land for housing: a political resource – reflections from Zimbabwe's urban areas. J South Afr Stud 41(6):1219–1238

Muchadenyika D, Williams JJ (2018) Politics, centralisation and service delivery in urban Zimbabwe. J South Afr Stud 44(5):833–853

Muchadenyika D, Williams JJ (2020) Central–local state contestations and urban management in Zimbabwe. J Contemp Afr Stud 38(1):89–102

Mukandavire Z, Liao S, Wang J, Gaff H, Smith DL, Morris JG Jr (2011) Estimating the reproductive numbers for the 2008–2009 cholera outbreaks in Zimbabwe. Proc Natl Acad Sci 108(21):8767–8772

Mukwazhi T (2015) Zimbabwe to street vendors: pack up, clean up, ship out! https://www.npr.org/sections/goatsandsoda/2015/06/15/414662878/zimbabwe-to-street-vendors-pack-up-clean-up-ship-out. Accessed 21 May 2022

Multi-Donor Trust Fund (MDTF) (2008) Zimbabwe Emergency Recovery Program, Draft 2, September, Harare

Munatsi B (2012) Zimbabwe: five deaths in Harare as typhoid continues to spread. https://www.humanitarianresponse.info/en/operations/zimbabwe/article/zimbabwe-five-deaths-harare-typhoid-continues-spread. Accessed 21 May 2022

Munzwa K, Wellington J (2010) Urban development in Zimbabwe: A human settlement perspective. Theor Empirical Res Urban Manage 5(5):120–146

Murray M, Myers G (eds) (2007) Cities in contemporary Africa. Springer, New York

Nangombe SS (2015) Drought conditions and management strategies in Zimbabwe. In Proceedings of the regional workshops on capacity development to support national drought management policies for eastern and southern Africa and the Near East and North Africa regions, pp 84–89

National Academies of Sciences, Engineering, and Medicine (2017) Addressing continuous threats: HIV/AIDS, Tuberculosis, and Malaria. Global health and the future role of the United States

News of the South (2017) Murambatsvina returns, council destroys 700 Harare South houses, August 18. https://newsofthesouth.com/murambatsvina-returns-council-destroys-700-harare-south-houses/. Accessed 21 May 2022

Nickol ME, Kindrachuk J (2019) A year of terror and a century of reflection: perspectives on the great influenza pandemic of 1918–1919. BMC Infect Dis 19:1–10

Pabst M (2001) Zimbabwe at the crossroads, Country report no. 3. South African Institute of International Affairs, Braamfontein

Pasqui G (2022) Coping with the pandemic in fragile cities. Springer Nature, Cham

Pettinger T (2019) Hyper inflation in Zimbabwe. https://www.economicshelp.org/blog/390/inflation/hyper-inflation-in-zimbabwe/. Accessed 24 May 2022

Polonsky JA, Martinez-Pino I, Nackers F, Chonzi P, Manangazira P, Van Herp M, Luquero FJ (2014) Descriptive epidemiology of typhoid fever during an epidemic in Harare, Zimbabwe, 2012. PLoS One 9(12):e114702

Poncin M, Marembo J, Sreenivasan N, Makwara I, Machekanyanga Z, Nyabyenda W, Manangazira P (2022) Implementation of an outbreak response vaccination campaign with typhoid conjugate vaccine–Harare, Zimbabwe, 2019

Potts D (2010) Internal migration in Zimbabwe: the impact of livelihood destruction in rural and urban areas. In: Zimbabwe's exodus: crisis, migration, survival. Southern African Migration Project, Waterloo, pp 79–111

Potts D (2011) 'We have a tiger by the tail': continuities and discontinuities in Zimbabwean city planning and politics. Crit Afr Stud 4(6):15–46

Potts D (2012) Whatever happened to Africa's rapid urbanization. Africa Research Institute, London

Raath J (2016) Mugabe fires 25,000 state workers. The Times. https://www.thetimes.co.uk/article/mugabe-fires-25-000-state-workers-tq8lcf0qz?region=global. Accessed 12 May 2022

Ratha D, Mohapatra S, Scheja E (2011) Impact of migration on economic and social development: a review of evidence and emerging issues. World Bank policy research working paper (5558)

Richardson CJ (2005) The loss of property rights and the collapse of Zimbabwe. Cato J 25:541

Rogerson CM (2016) Responding to informality in urban Africa: street trading in Harare, Zimbabwe. Urban Forum 27:229–251

Runganga R, Njoroge W, Mishi S (2022) Restoration of land acquired for resettlement and the Fast-track land reform programme in Zimbabwe. Sustainability 14(15):9178

Sachikonye LM (2003) The situation of commercial farm workers after land reform in Zimbabwe. Farm Community Trust of Zimbabwe, Harare

Samaita K (2019) Zimbabwe's inflation doubles up to 175%. BusinessLIVE. https://www.businesslive.co.za/bd/world/africa/2019-07-15-zimbabwes-inflation-doubles-up-to-175/. Accessed 21 May 2022

Skovdal M, Campbell C, Nyamukapa C, Gregson S (2011) When masculinity interferes with women's treatment of HIV infection: a qualitative study about adherence to antiretroviral therapy in Zimbabwe. J Int AIDS Soc 14(1):1–7

Snyder RE, Boone CE, Cardoso CAA, Aguiar-Alves F, Neves FP, Riley LW (2017) Zika: a scourge in urban slums. PLoS Negl Trop Dis 11(3):e0005287

Steele AD, Hay Burgess DC, Diaz Z, Carey ME, Zaidi AK (2016) Challenges and opportunities for typhoid fever control: a call for coordinated action. Clin Infect Dis 62(suppl_1):S4–S8

Stiff P (2000) Cry Zimbabwe: independence – twenty years on. Galago Publishing, Johannesburg

Swanson MW (1977) The sanitation syndrome: bubonic plague and urban native policy in the Cape colony, 1900–19091. J Afr Hist 18(3):387–410

Terceira N, Gregson S, Zaba B, Mason P (2003) The contribution of HIV to fertility decline in rural Zimbabwe, 1985–2000. Popul Stud 57(2):149–164

The Herald (2015) 200 houses demolished, June 18. https://www.herald.co.zw/200-houses-demolished/. Accessed 21 May 2022

The Herald (2018) Zimbabwe has the world's largest informal economy: IMF, May 19. https://www.herald.co.zw/zim-has-worlds-second-largest-informal-economy-imf/. Accessed 22 May 2022

The Standard (2016) Why property is more pricey in Zim than SA. https://www.thestandard.co.zw/2016/06/27/property-pricey-sa-zim. Accessed 28 June 2022

The Zimbabwe Mail (2020) Harare mayor halts informal market demolitions after threats from MDC, July 15. https://www.thezimbabwemail.com/business/harare-mayor-halts-informal-market-demolitions-after-threats-from-mdc/. Accessed 28 June 2022

Tibaijuka AK (2005) Report of the fact-finding mission to Zimbabwe to assess the scope and impact of operation murambatsvina by the UN special envoy on human settlements issues in Zimbabwe. United Nations, New York

Tlhajoane M, Masoka T, Mpandaguta E, Rhead R, Church K, Wringe A, Gregson S (2018) A longitudinal review of national HIV policy and progress made in health facility implementation in Eastern Zimbabwe. Health Res Policy Syst 16(1):1–13

Toriro P, Chirisa I (2021) Vendors on wheels! The changing terrain and manifestation of informality in Harare under Covid-19 pandemic restrictions. Cogent Soc Sci 7(1):1939230

Turok I (2014) Linking urbanisation and development in Africa's economic revival. Zed Books, London, pp 60–81

UN-Habitat (2017) New Urban Agenda, sustainable cities take spotlight at General Assembly event. https://www.un.org/sustainabledevelopment/blog/2017/09/new-urban-agenda-reform-of-un-habitat-take-spotlight-athigh-level-general-assembly-talks/. Accessed 3 Jan 2023

United Nations (2014) World urbanisation prospects: the 2014 revision, highlights. Department of Economic and Social Affairs, Population Division. https://www.un.org/en/development/desa/publications/2014-revision-worldurbanization-prospects.html. Accessed 23 May 2022

United Nations (2015) World urbanization prospects. Revision. Department of Economic and Social Affairs. http://www.un.org/en/development/desa/publications/2018-revision-world-urbanization-prospects.html. Accessed 23 May 2022

UNAIDS (2020) Zimbabwe. https://www.unaids.org/en/regionscountries/countries/zimbabwe. Accessed 20 May 2022

United Nations (2018) World urbanization prospects 2018. https://population.un.org/wup/. Accessed 21 July 2022

United Nations World Population Prospects (2018) https://www.un.org/en/desa/2018-revision-world-urbanizationprospects. 2018 Revision of World Urbanization Prospects. Accessed 23 May 2022

United Nations, Department of Economic and Social Affairs (2014) World urbanization prospects. Revision. http://www.un.org/en/development/desa/publications/2018-revision-world-urbanization-prospects.html. Accessed 23 May 2022

United Nations, Department of Economic and Social Affairs, Population Division (2014) World urbanisation prospects: the 2014 revision, highlights

van Gelder E (2008) Zimbabwe: black market thriving. Africa News

Venables N (2020) Finding home in uncertainty: returnees, reintegration, and reconciliation a case study of refugees in towns Harare, Zimbabwe. Feinstein International Center. https://fic.tufts.edu/wp-content/uploads/Zimbabwe.pdf. Accessed 4 Feb 2023

WHO (2016) Urbanisation and Health. Bull World Health Organ. World Health Organization. http://www.who.int/bulletin/volumes/88/4/10-010410/en/. Accessed 3 Jan 2023

World Bank (2020) COVID-19 to add as many as 150 million extreme poor by 2021, October 7. https://www.worldbank.org/en/news/press-release/2020/10/07/covid-19-to-add-as-many-as-150-million-extreme-poor-by-2021. Accessed 10 Feb 2021

World Health Organization (2018) Typhoid vaccines. Weekly epidemiological record, WHO position paper 93(13). https://www.who.int/publications-detail-redirect/typhoid-vaccines-who-position-paper-march-2018. Accessed 4 Feb 2023

World Health Organization (2019) Africa's first-ever mass typhoid fever vaccination campaign ends in Zimbabwe, March 6. https://www.afro.who.int/news/africas-first-ever-mass-typhoid-fever-vaccination-campaign-ends-zimbabwe. Accessed 4 Feb 2023

World Health Organization (n.d.) Urbanisation and Health. Bull World Health Organ. http://www.who.int/bulletin/volumes/88/4/10-010410/en/. Accessed 3 Jan 2023

Writer S (2014) A look at Zimbabwe's property market. BusinessTech. https://businesstech.co.za/news/property/310256/a-look-at-zimbabwes-property-market/. Accessed 4 Feb 2023

Yoshikuni T (2007) African urban experiences in colonial Zimbabwe; a social history of Harare before 1925. Weaver Press, Harare

Zeiderman A, Kaker S, Silver J, Wood A, Ramakrishnan K (2017) Urban uncertainty: governing cities in turbulent times. LSE Cities, New York

Zinyama L, Tevera D (2017) Zimbabwe: in search of spatial and social equity. In: Geography and economy in South Africa and its neighbours. Routledge, London, pp 231–251

Zinyama LM, Tevera DS, Cumming SD (eds) (1993) Harare: the growth and problems of the city. University of Zimbabwe Publications, Harare

Chapter 4
The COVID-19 Pandemic and Urban Policy Interventions in Zimbabwe

Introduction

The first official case of COVID-19 in Zimbabwe was recorded on Friday, March 20, 2020, from a citizen who had travelled abroad and was in Victoria Falls at the time (The Herald Zimbabwe 2021). Initially, incidences of COVID-19 were reported from people who had travelled to countries with high numbers of cases (Murewanhema et al. 2020). However, cases began to increase as the virus spread within the country, especially in urban areas. This increase was attributed to many factors, including population density, lack of awareness about the virus, and limited access to testing and healthcare (Kollamparambil and Oyenubi 2021). As of 2021, there was a surge in cases, and the COVID-19 pandemic significantly impacted urban policy interventions in Zimbabwe and many other countries worldwide (Chirisa et al. 2021).

In response to the increase in COVID-19 cases, the Zimbabwean government introduced policies to curb the spread of the virus (Haider et al. 2020). These included implementing a nationwide lockdown, closing of borders, and suspending public gatherings (Haider et al. 2020). They also instituted mandatory face masks in public spaces, social distancing, and increased testing and contact tracing, to mention a few (Gkiotsalitis and Cats 2021). These measures were implemented to slow the virus' spread and protect the health and safety of the population. Additionally, the government raised awareness about the virus and its spread through public education campaigns (Bhanye and Bhanye 2023; Dzobo et al. 2020). While the primary goal of these policies was to curb the spread of the virus, they were shaped by political, economic, or social considerations. According to Fisher et al. (2020), it is important to note that the policies and interventions implemented to address the pandemic impacted different individuals and groups within a society. For example, certain policies, such as mandatory lockdowns or travel restrictions,

J. I. Bhanye et al., *COVID-19 Lockdowns and the Urban Poor in Harare, Zimbabwe*, https://doi.org/10.1007/978-3-031-41669-9_4

disproportionately affected marginalized or low-income communities. It is also worth considering how the pandemic and its response may exacerbate societal inequalities.

Urban policy interventions were critical in addressing the COVID-19 pandemic in Zimbabwe (Chirisa et al. 2022). By implementing measures to slow the spread of the virus and provide support to those affected, cities and local governments were able to mitigate the negative impacts of the pandemic on their communities (Mackworth-Young et al. 2021). However, the nature of the pandemic continued to pose challenges for urban areas, calling for governments and city authorities to devise strategic measures to adapt and respond to the pandemic and protect the health and well-being of citizens, especially in vulnerable urban spaces.

Therefore, it is essential to critically evaluate the policies and interventions implemented to address the COVID-19 pandemic, not only in terms of their effectiveness in controlling the spread of the virus but also in terms of their impact on different individuals and groups within a society (Güner et al. 2020). This chapter begins by articulating the policy perspectives from an urban planning perspective, discussing the different policy orientations. Next, the chapter zooms into the COVID-19 policy interventions adopted in Zimbabwe. Policies are decisions by governments on what actions to take or not take to address specific issues or problems. In the case of the COVID-19 pandemic, governments around the world implemented a range of policies and interventions, such as lockdowns, travel restrictions, testing and contact tracing, and vaccination campaigns, to curb the spread of the virus and protect public health (Park and Ranabhat 2022). Weible et al. (2020) mentioned that the policies are not neutral. They are shaped by the ideologies, values, and priorities of those who design and implement them.

The COVID-19 pandemic had a significant impact on urban policy in Zimbabwe and countries around the world, as governments and local authorities had to rapidly adapt and implement new measures to contain the spread of the virus and mitigate its effects on the population and economy (World Bank 2021a, b). The pandemic highlighted the need for more comprehensive and resilient urban policies that are better prepared to respond to future crises. Therefore, it is important to critically evaluate the policies and interventions implemented to address the pandemic, not only in terms of their effectiveness in controlling the spread of the virus, but also in terms of their impact on different individuals and groups within a society. Therefore, the following subheading will focus on policymaking in cities under crisis.

Policymaking in Cities Under Crisis

Policymaking can be complex and challenging in cities facing a crisis (OECD 2017). Many factors are considered when developing policies in these situations, including the nature of the crisis, resources available, and the community's needs. One of the key considerations when developing policies in a crisis is the need to prioritize the safety and well-being of the community. This involves addressing

Fig. 4.1 Policymaking approaches for cities under crisis. (Source: Authors 2023)

immediate threats to public health, such as a natural disaster or a disease outbreak, or addressing long-term challenges such as economic or social inequality (Gilmore et al. 2020). Effective policymaking in a crisis requires effective communication and collaboration between different levels of government, community organizations, and other stakeholders (Krause et al. 2020; Matamanda et al. 2022). This can ensure that policies the developed are transparent and inclusive of the community's needs and concerns (Hyland-Wood et al. 2021). Several policy approaches can be taken when developing policies in a crisis, as shown in Fig. 4.1.

Rational Comprehensive Policy

The Hungarian–American planner and economist Alexander R. Faludi developed the rational comprehensive policymaking approach. This theory suggests that policymaking should be a rational, systematic, and comprehensive approach to problem-solving (Roose and Kull 2012). The theory holds that government officials, experts, and other stakeholders should use scientific methods, data, and expert knowledge to identify and analyze problems and then design and implement policies likely to address those problems. Proponents of this theory argue that this approach leads to well-informed, logical, and practical policy decisions that can be justified by sound evidence and expert analysis. They also argue that this approach is more efficient than other methods, as it allows for more effective use of resources and reduces unintended consequences.

Rational comprehensive policymaking involves a thorough analysis of all the options and alternatives and a systematic evaluation of the costs and benefits of each option (Uzonwanne 2016). This approach to policymaking is that rational, evidence-based decisions can lead to the best outcomes for a community or society (Cairney and Oliver 2017). In the context of cities facing a crisis such as a natural disaster, economic downturn, or public health emergency, the rational comprehensive approach can be instrumental (Harrison and Williams 2016). This is because it allows policymakers to carefully consider the long-term impacts of their decisions and systematically weigh the costs and benefits of different options. To implement a rational comprehensive approach to policymaking in a city facing a crisis, policymakers should follow the steps articulated in Fig. 4.2.

However, critics argue that the rational comprehensive approach to policymaking is unrealistic, as it does not consider the complexity of real-world problems,

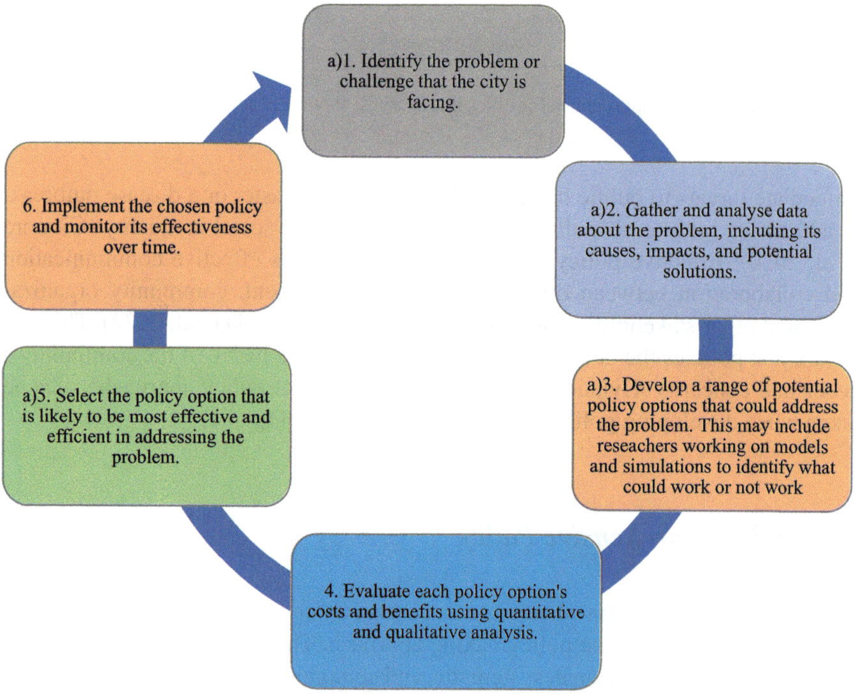

Fig. 4.2 Rational comprehensive policymaking approach. (Source: Authors)

limitation of human cognitive abilities, and the policymaking process political realities (Forester 1984). They argue that it is unrealistic to expect that all stakeholders will agree on a policy's facts, values, and objectives and that it is unlikely that all relevant information will be available at the time of decision-making. In reality, policymaking is often far from rational and comprehensive and is influenced by various factors, including politics, power, and the interests of stakeholders (Kørnøv and Thissen 2000).

Proponents of the rational comprehensive approach to policymaking argue that it is a logical and systematic way to solve problems. However, it is important to note that the rigidity of this approach and its obsession with theory can be a limitation (Turner 2016). The approach assumes that policymaking can be perfectly rational and that all stakeholders agree on a policy's facts, values, and objectives. However, in reality, policymaking is influenced by various factors such as politics, power, and the interests of different stakeholders. This can lead to a lack of flexibility and an inability to adapt to changing circumstances (Hudson et al. 2019). Additionally, the obsession with theory can lead to a lack of attention to practical considerations and the real-world impact of policies (Lapuente and Van de Walle 2020). The focus on theoretical models and abstract concepts can lead to policies not well-suited to the specific context in which they are implemented and the needs and perspectives of those affected by the policy are not considered.

One example of how the rational comprehensive approach to policymaking may be applied to a problem like climate change is through the United Nations Framework Convention on Climate Change and the Conference of the Parties meetings. The Conference of the Parties meetings bring together policymakers, scientists, and other stakeholders worldwide to negotiate and implement policies to address climate change. The process follows a rational and systematic approach, aiming to identify and analyze the problem of climate change and then design and implement policies likely to address it effectively. However, negotiating and implementing climate change policies is not straightforward, and it takes time to achieve consensus among all the countries and stakeholders. The cost implications of climate finance, such as the cost of transitioning to renewable energy and the cost of adapting to the impacts of climate change, are significant and have to be considered when making decisions (Bhandary et al. 2021).

The rational comprehensive approach to policymaking may be appropriate for problems like climate change that evolve. However, it is important to recognize that negotiating and implementing climate change policies can be complex and take time and that the cost implications of climate finance must be considered (Orlove et al. 2020). Other approaches to policymaking could also be applied to address climate change, such as adaptive governance, incrementalism, and the advocacy coalition framework. The rational comprehensive approach is not the only policymaking model and may not always be the most appropriate or effective in every situation. Other approaches, such as incremental or participatory policymaking, may also be used in certain circumstances (Kuruvilla and Dorstewitz 2010).

Incremental Planning Policy

Charles Lindbloom postulated the incremental planning policymaking approach. In his conceptualization of this approach, Lindblom (1959) critiqued the comprehensive rational model arguing that it was only ideal for simple problems. Lindbloom postulated that instead of the linear approach adopted for rational comprehensive policies, complex problems, including crises such as the COVID-19 pandemic, require muddling through. This involves making small, gradual changes to existing policies and plans over time rather than making major, sweeping changes all at once. A key element of the muddling-through policy approach is its practicality. It allows decision-makers to be practical and deal with the matter at hand without being too engrossed in theories as applied in the comprehensive rational model.

Rydin et al. (2012) commented that this approach is used when there is a need to address pressing issues or challenges carefully and accurately. In the context of cities under crisis, incremental planning policy might involve making minor changes to existing policies and plans to address specific challenges or issues that have emerged due to the crisis (Lindblom 1959; UN-Habitat 2020). This involves adjusting existing policies and plans to address immediate needs, such as providing emergency housing or other forms of support for vulnerable populations, while also considering longer-term solutions that can address the underlying causes of the crisis. For example, during a cholera outbreak, this would involve adjusting existing policies and plans to address immediate needs, such as providing emergency housing or support for vulnerable populations, while also considering longer-term solutions to address the underlying causes of the outbreak. Some of the specific policy interventions that might be included in an incremental planning approach in a city under crisis include those illustrated in Box 4.1.

Box 4.1: Policy Interventions That Might Be Included in an Incremental Planning Approach

(a) *Emergency housing and support:* In a crisis, such as a natural disaster or a pandemic, incremental planning policy might involve the provision of emergency housing and other forms of support for vulnerable populations. This could include temporary shelters, food and other necessities, and other forms of immediate assistance. Examples have been seen in how governments have reacted in times of crisis, like Hurricane Katrina in New Orleans, USA (Barnes et al. 2008). The same was also noted in Zimbabwe, Mozambique, and Zambia during Cyclone Idai, ravaging the region (Charrua et al. 2021).

(b) *Short-term solutions:* Incremental planning policy might also involve implementing short-term solutions to address specific challenges or issues resulting from the crisis. These solutions might include temporary measures to provide essential services or support to affected communi-

(continued)

Box 4.1 (continued)

ties, such as providing temporary healthcare facilities or establishing food banks. For example, during the cholera outbreak of 2008/2009 in Zimbabwe that claimed at least 4000 lives, the government was proactive in providing temporary health facilities (Cuneo et al. 2017).

(c) *Longer-term solutions:* In addition to addressing immediate needs, incremental planning policy might also involve considering longer-term solutions that can address the underlying causes of the crisis. These might include policies and initiatives aiming to improve the community's resilience or address structural issues contributing to vulnerability. The same was also noted in West Africa during the Ebola pandemic, where governments and NGOs provided long-term solutions such as community education, awareness, and upscaling risk communication (Buseh et al., 2015).

Figure 4.3 shows a summary of the policy interventions in incremental planning during a complex problem or crisis. Policy responses (short-term response and long-term policy interventions) and the specific policy actions (e.g., housing provision, medical assistance, and food supplies for short-term immediate response and investing in public health infrastructure for long-term policy interventions) are illustrated.

Incremental planning is a policymaking approach that involves making small, gradual changes to existing policies and plans over time rather than making significant, sweeping changes all at once (Bennett and Howlett 1992). This approach is used when there is a need to address pressing issues or challenges in a careful and measured way. It can be particularly useful in the context of cities under crisis.

Strategic Planning Policies

Strategic planning policies are long-term plans that outline an organization or community's goals and objectives and the strategies and actions that will be taken to achieve those goals. These policies are designed to guide decision-making and

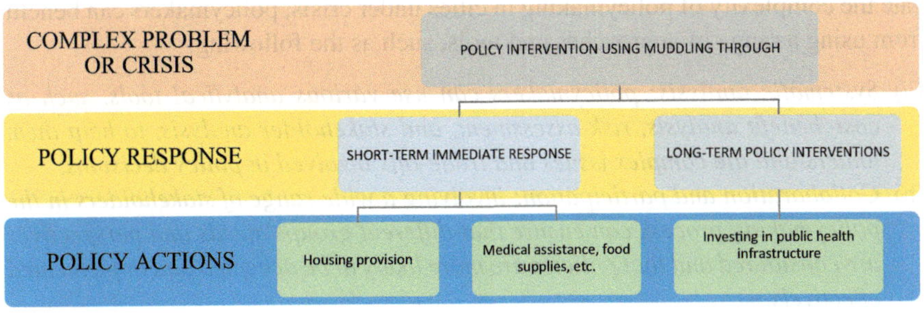

Fig. 4.3 Summary of the policy interventions in incremental planning. (Source: Authors)

ensure that resources are used effectively and efficiently to achieve the desired outcomes. In the context of cities facing a crisis, such as a natural disaster, economic downturn, or public health emergency, strategic planning policies can be particularly useful in guiding the development and implementation of response and recovery efforts (WHO 2017). These policies can ensure that the city's resources are used in a coordinated and targeted way to address the most pressing needs and priorities and to support the long-term resilience and sustainability of the community. To develop and implement strategic planning policies in a city facing a crisis, policymakers should:

(a) *Identify the goals and objectives the city wants to achieve in response to the crisis.*
(b) *Conduct a thorough analysis of the current situation and the resources available to the city.*
(c) *Develop a range of strategies and actions that could help achieve the desired goals and objectives.*
(d) *Evaluate the costs and benefits of each strategy and action using quantitative and qualitative analysis.*
(e) *Select the strategies and actions that are likely to be most effective and efficient in addressing the crisis.*
(f) *Implement the chosen strategies and actions, and monitor their effectiveness over time.*

It is important that the development and implementation of strategic planning policies may involve input and participation from a wide range of stakeholders, including government agencies, community organizations, and members of the public.

Complexity

Policymaking can be particularly complex in cities facing a crisis due to the range of challenges and issues that must be addressed and the diverse interests and perspectives of the stakeholders involved (OECD 2017). In addition, the rapidly changing nature of many crises can make it challenging to predict policy decisions' long-term impacts and develop effective strategies and actions. To effectively navigate the complexity of policymaking in cities under crisis, policymakers can benefit from using a range of approaches and tools, such as the following:

(a) *Systematic analysis: policymakers can use various analytical tools, such as cost–benefit analysis, risk assessment, and stakeholder analysis, to help them understand the complex issues and trade-offs involved in policy decisions.*
(b) *Collaboration and participation: involving a wide range of stakeholders in the policymaking process can ensure that different groups' needs and perspectives are considered and that policies are more likely to be accepted and implemented effectively.*

(c) *Adaptive management: policymakers can use an adaptive management approach, which involves continuous monitoring and adjusting policies and actions in response to changing circumstances, to respond effectively to many crises' complex and uncertain natures.*
(d) *Planning and scenario analysis: policymakers can use planning and scenario analysis tools to anticipate and prepare for a range of potential future developments and to develop contingency plans in case of unexpected events.*

Using these approaches and tools, policymakers can better understand and address the complexity of policymaking in cities under crisis. They can make informed and effective decisions to support the resilience and well-being of their communities. Effective policymaking in a crisis requires a combination of strong leadership, effective communication, and a commitment to addressing the needs and concerns of the community (Hyland-Wood et al. 2021).

Ideologies Behind Policies

Policies are often based on ideologies or philosophical beliefs about the role and function of government, the economy, and society. These ideologies shape the values, priorities, and goals that drive policy decisions and influence development and implementation (Harrison and Boyd 2018). Some common ideologies that shape policymaking are discussed in the following sections.

Socialist Policies

Socialist policies have their roots in socialist countries where equity is advocated. The idea is that the economy and society should be owned and controlled by the community rather than private individuals or corporations (Steinmetz 2019). In socialist societies, the government plays a central role in managing the economy and redistributing wealth and resources to promote social and economic equality. The core issue is to ensure socioeconomic equity among the citizens regardless of race, religion, age, or ethnic background. In the early 1980s, both Cuba and Zimbabwe were ruled by socialist governments, which implemented urban policies based on socialist principles (Rwodzi and Fosl 2020). These policies focused on providing essential services, such as housing, healthcare, and education, to all members of society, regardless of their income or social status. However, several proponents argue that these are populist ideologies that are often rhetoric, as evident in Zimbabwe (de Vreese et al. 2018).

The Cuban government implemented urban policies to improve the population's living conditions, including constructing new housing developments and expanding healthcare and education facilities (Luzon 1988). The government also implemented policies to promote social and economic equality, such as land reform programs and the nationalization of particular industries. Similarly, in Zimbabwe, the government implemented a range of urban policies to improve the population's living conditions, including the construction of new housing developments, the expansion of healthcare and education facilities, and the nationalization of specific industries. The Government of Zimbabwe implemented policies to promote social and economic equality, such as land reform programs and the establishment of cooperatives. Socialist policies in Cuba and Zimbabwe in the early 1980s promoted social and economic equality to improve the population's living conditions by providing essential services and redistributing wealth and resources (Cook and Yi 2011).

Neoliberal Policies

According to Navarro (2007), the idea of neoliberal policies is that the market should be the primary mechanism for organizing economic and social activity. These policies prioritize economic growth, individual freedom, and free trade and advocate a minor role for the state in the economy and society (Carlquist and Phelps 2014). In the urban context, neoliberal policies privatize public services and resources, degrade the economy, and promote market-based solutions to urban problems (Harvey 2007). These policies are typically justified because they promote efficiency, innovation, and economic growth. Some of the specific urban policy interventions that are associated with neoliberalism include the following:

(a) *Privatization: neoliberal policies privatize public services and resources, such as water and sanitation, housing, and transportation. This involves the transfer of ownership and control of these assets from the public sector to the private sector.*
(b) *Deregulation: neoliberal policies deregulate the economy, including removing or reducing regulations and restrictions on business and industry. This includes relaxing zoning and building regulations, removing tariffs and other trade barriers, and reducing labor protections.*
(c) *Market-based solutions: neoliberal policies promote market-based solutions to urban problems, such as using tolls and user fees to fund transportation infrastructure or using public–private partnerships to finance and manage public services and resources.*

The idea of neoliberal policies in the urban context is that the market should be the primary mechanism for organizing economic and social activity and prioritize economic growth, individual freedom, and free trade (Haque 1999). These policies privatize public services and resources, deregulate the economy, and promote market-based solutions to urban problems.

Capitalist Policies

Capitalist policies promulgate that the economy should be driven by private enterprise and the pursuit of profit (Bowles 2007). These policies prioritize the rights of private property owners and the freedom of individuals and businesses to engage in market transactions. In the urban context, capitalist policies involve promoting a market-based approach to land use and development, relying on private enterprises to provide goods and services, and use market mechanisms to allocate resources and distribute wealth. Some of the urban policy interventions associated with capitalism are as follows:

(a) *Land use and development: capitalist policies involve a market-based approach to land use and development, with private property owners having the right to use and develop their land as they see fit, subject to certain regulatory constraints. This includes property taxes and other market-based mechanisms to finance public goods and services.*
(b) *Private enterprise: capitalist policies rely on private enterprises to provide goods and services, such as housing, transportation, and utilities. This involves the privatization of public assets and using private-sector contractors to deliver public services.*
(c) *Market mechanisms: capitalist policies rely on market mechanisms to allocate resources and distribute wealth, including using prices and other economic signals to guide the allocation of resources. This involves using market-based approaches to finance and manage public goods and services, such as user fees or tolls.*

The idea of capitalist policies in the urban is that the economy should be driven by private enterprise, the pursuit of profit, prioritize the rights of private property owners, and the freedom of individuals and businesses to engage in market transactions (Gatwiri et al. 2020). These policies involve promoting a market-based approach to land use and development, relying on private enterprises to provide goods and services, and using market mechanisms to allocate resources and distribute wealth.

People-Centered Policies

People-centered policies are designed to prioritize the needs and well-being of individuals and communities (Santana et al. 2018). These policies focus on meeting the population's basic needs, such as access to housing, healthcare, education, and other essential services. In the urban context, people-centered policies involve providing affordable housing, expanding public transportation and other infrastructure, and promoting access to public spaces and amenities (Hofer et al. 2022). These policies are justified because they promote social and economic inclusion and reduce inequality. Some of the urban policy interventions that are associated with people-centered policies include the following:

(a) *Affordable housing: people-centered policies provide affordable housing options for low-income households, such as social housing programs. These programs are designed to ensure that all community members can access safe and affordable housing.*
(b) *Public transportation and infrastructure: people-centered policies involve the expansion of public transportation and other infrastructure, such as roads, sidewalks, and bicycle lanes, to ensure that all members of the community have access to these essential services.*
(c) *Public spaces and amenities: people-centered policies promote access to public spaces and amenities, such as parks, playgrounds, and community centers, to ensure that all community members have access to these resources.*

People-centered policies in the urban context prioritize the needs and well-being of individuals and communities. These policies focus on meeting the population's basic needs, such as access to housing, health are, education, and other essential services. They are justified because they promote social and economic inclusion and reduce inequality.

Bottom-Up/Top-Down Policies

Bottom-up and top-down policies refer to how decisions about policy and planning are made and implemented (Cerna 2014). Bottom-up policies are developed and implemented from the ground up, with input and participation from individuals and communities (Koontz and Newig 2014). These policies are based on the idea that local communities and stakeholders are best placed to identify and address their own needs and priorities. Top-down policies, on the other hand, are developed and implemented by higher levels of government or other authorities, with little or no input or participation from local communities. These policies are based on the idea that higher levels of government or other authorities are better equipped to make decisions about policy and planning and that local communities should implement these policies (Hudson et al. 2019). In the urban context, bottom-up policies include community-led planning processes, participatory budgeting, or citizen-led advocacy campaigns. Top-down policies include decisions made by higher levels of government or other authorities about land use and development, transportation, or infrastructure investment, without the input or participation of local communities (Horn et al. 2018).

Interventionist Policies

Interventionist policies involve the active intervention of the state or other authorities in the economy and society to achieve specific goals or outcomes. These policies are based on the idea that the state or other authorities have a role in addressing social and economic problems (Van der Waldt 2015) and that intervention is necessary to achieve particular objectives. In the urban context, interventionist policies include public housing programs, transportation planning and investment, or the provision of infrastructure and public amenities. These policies are justified because they promote social and economic development, reduce inequality, or address other pressing urban issues (Scheba et al. 2021). Some of the urban policy interventions that are associated with interventionist policies include the following:

(a) *Public housing programs: interventionist policies involve the provision of public housing for low-income households, such as social housing. These programs are designed to ensure that all community members can access safe and affordable housing.*
(b) *Transportation planning and investment: interventionist policies involve active planning and investment in transportation infrastructure and services, such as public transit, roads, and pedestrian and bicycle infrastructure. These policies are designed to ensure that all community members have access to reliable and affordable transportation.*
(c) *Infrastructure and public amenities: interventionist policies involve the provision of infrastructure and public amenities, such as parks, playgrounds, and community centers, to ensure that all members of the community have access to these resources.*

The interventionist policies in the urban context involve the active intervention of the state or other authorities in the economy and society to achieve specific goals or outcomes. These policies are justified because they promote social and economic development, reduce inequality, or address other pressing urban issues (Cingano 2014). Many other ideologies can shape policymaking, and it is common for policies to reflect a mix of different ideological influences. Understanding the ideologies behind policies provides context and insight into the motivations and goals of policymakers (Cerna 2014). They inform discussions about the impacts and implications of different policy options.

The Rationale for COVID-19 Policy Interventions

World Health Organization's Role in the Policy Terrain

The WHO is a specialized agency of the United Nations responsible for providing global leadership on public health matters. As such, the WHO played a vital role in shaping policy response to the COVID-19 pandemic (Guo et al. 2022). In the context of the COVID-19 pandemic, the WHO provided guidance and recommendations on a range of policy interventions to reduce the spread of the virus and protect public health (Güner et al. 2020). These recommendations were based on the best available scientific evidence and designed to help countries worldwide respond to the pandemic. Some examples of policy recommendations made by the WHO in the context of COVID-19 include:

(a) *Implementing measures to restrict movement and reduce social interaction, such as lockdowns and mask mandates.*
(b) *Promoting hand hygiene and physical distancing as effective ways to reduce the transmission of the virus.*
(c) *Providing financial and other forms of support to businesses and individuals affected by the pandemic.*
(d) *Enhancing surveillance and testing capacity to better track and understand the spread of the virus.*
(e) *Supporting the development and deployment of vaccines and other tools to prevent and treat COVID-19.*

In addition to providing guidance and recommendations, the WHO played a role in coordinating the global response to the pandemic. This included working with member states to share information and best practices and providing technical assistance and support to countries struggling to contain the spread of the virus (WHO 2020). Overall, the WHO played a vital role in shaping the policy response to the COVID-19 pandemic, both through developing and disseminating guidance and recommendations and by coordinating the global response (Anwar et al. 2020).

Southern African Development Community (SADC) Policy Contributions

The SADC[1] member states implemented a range of policy interventions to respond to the COVID-19 pandemic, including measures to prevent the spread of the virus, support the healthcare system, and mitigate the economic and social impacts of the pandemic (OCHA 2021). One of the critical policy interventions implemented by SADC was implementing public health measures such as lockdowns, travel restrictions, and the mandatory use of face masks in public places. Other measures implemented included the closure of schools and nonessential businesses, the implementation of mask mandates, and the promotion of hand hygiene and physical distancing. These measures were put in place to reduce the transmission of the virus and protect the population, particularly the most vulnerable (Ayouni et al. 2021).

[1] The Southern African Development Community (SADC) is a regional intergovernmental organization comprising 15 member states in Southern Africa.

In addition to public health measures, SADC implemented economic and social policies to support individuals and businesses affected by the pandemic (Rosenberg et al. 2021). Policies included financial support for businesses and workers affected by the pandemic, tax relief, emergency funding to support the healthcare system, and initiatives to support economic recovery and stimulate economic growth (OECD 2020a, b, c, d). The policy interventions implemented by SADC in response to the COVID-19 pandemic were designed to protect public health, support individuals and businesses affected by the pandemic, and mitigate the economic and social impacts of the crisis (Nicola et al. 2020).

Zimbabwe National Government and COVID-19

The Government of Zimbabwe implemented many policy interventions in response to the COVID-19 pandemic (Chitungo et al. 2022). Interventions were implemented to slow the virus' spread and protect the population's health and well-being. In March 2020, when the first cases of COVID-19 were reported in the country, the government quickly implemented measures to curb the spread of the virus (Maulani et al. 2020). These measures included travel restrictions, the closure of nonessential businesses, and the imposition of a nationwide curfew. In addition, the government set up a National COVID-19 Taskforce to coordinate the country's response to the pandemic (Masiya et al. 2021). The task force comprised representatives from government departments and agencies, experts from the private sector, and civil society. The task force was responsible for developing and implementing policies and strategies to combat the spread of COVID-19.

The Zimbabwean government increased the country's capacity to test for COVID-19. In May 2020, the government launched a national testing campaign to test at least 10% of the population for COVID-19 (WHO 2021). Several testing centers were established nationwide, including rural areas, to make testing more accessible to citizens. The government also played an important role in public health interventions. They provided public awareness campaigns to inform citizens about the virus and how to protect themselves (Mutanda 2022). Personal protective equipment was provided to healthcare and essential workers.

The government supported the country's economic recovery, which was significantly influenced by the pandemic. Economic relief measures were provided and included tax relief to businesses affected by the pandemic. Financial assistance was given to individuals and households affected by the pandemic. The government ensured the distribution and accessibility of vaccines to the population and partnered with the WHO and UNICEF to secure vaccines (Murewanhema et al. 2022). Nationwide vaccination centers were established to make vaccines more accessible to citizens.

Zimbabwe, like many other countries, implemented a vaccination program in response to the COVID-19 pandemic. The country approved the use of several vaccines, including Sinopharm, Sinovac, Covaxin®, and Johnson and Johnson vaccines.

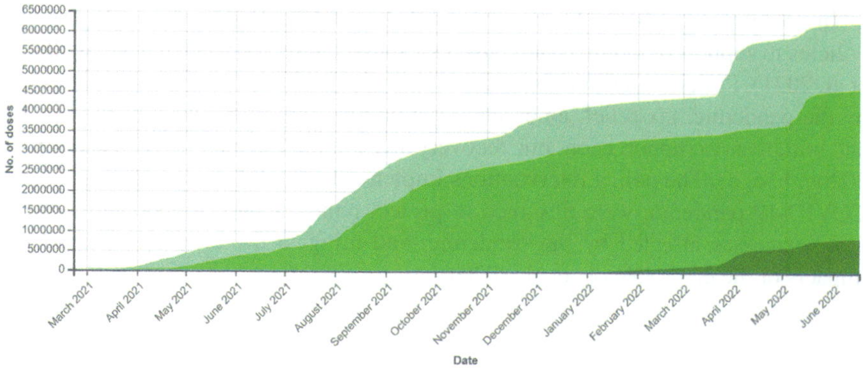

Fig. 4.4 Cumulative COVID-19 vaccine doses administered in Zimbabwe by June 2022. (Source: Ministry of Health and Childcare Zimbabwe 2022)

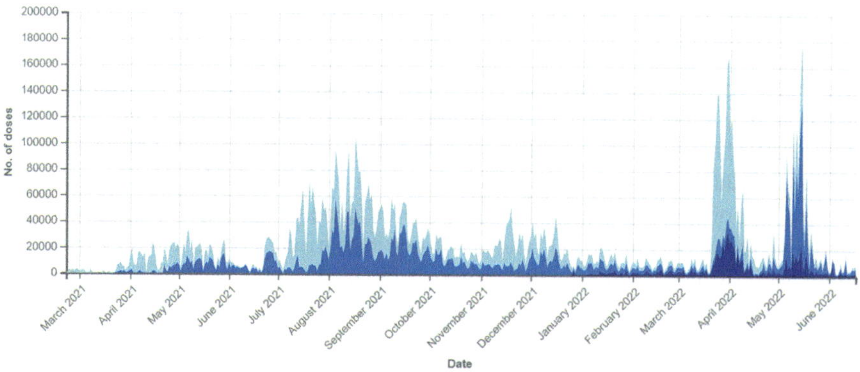

Fig. 4.5 Daily doses administered in Zimbabwe by June 2022. (Source: Ministry of Health and Childcare Zimbabwe 2022)

The vaccination program in Zimbabwe began in February 2021, with priority given to healthcare workers, the elderly, and those with underlying health conditions. The government launched a nationwide vaccination campaign to encourage as many people as possible to get vaccinated. As of June 17, 2022, 6,260,228 people received their first dose, 4,598,703 received their second dose, and 851,874 received a third dose (MoHCCZim 2022). Figure 4.4 shows cumulative COVID-19 vaccine doses administered in Zimbabwe by 2022; while Fig. 4.5 shows daily doses administered in Zimbabwe by June 2022.

Corruption was also reported within the public vaccination program, with those willing to pay bribes to hospital staff and members of Zimbabwe's ruling party ZANU–PF given priority. Vaccines were reportedly available within the private healthcare system at a cost of approximately US$40, automatically eliminating the urban poor especially during the early stages of the introduction of vaccines (York 2021). Later, the Government of Zimbabwe implemented policies to make the vaccine more accessible to people, including mobile vaccination clinics in remote areas and working with community leaders to raise awareness about the importance of vaccination. Vaccine hesitancy was a big issue in Zimbabwe, just like several other countries with people reluctant to get vaccinated due to concerns about the safety and effectiveness of the vaccine driven by the infodemic.[2] The government addressed these concerns and increased vaccine uptake by providing accurate information about the vaccines and their benefits.

Overall, the Zimbabwean government took a comprehensive approach to tackle the COVID-19 pandemic. The government's efforts focused on controlling the spread of the virus, supporting citizens and businesses affected by the pandemic, and increasing the country's capacity to test and vaccinate citizens. The policy interventions implemented by the Government of Zimbabwe were designed to protect the health and well-being of the population, slow the spread of the virus, and mitigate the economic and social impacts of the pandemic (World Bank 2021a, b).

COVID-19 Statistics in Zimbabwe

According to the Ministry of Health and Child Care Zimbabwe (2022), as of July 2022, Zimbabwe reported 256,047 infections and 5575 coronavirus-related deaths since the pandemic began. The country implemented various measures to contain the spread of the virus, including lockdowns, quarantine measures, and restrictions on travel and public gatherings (Bhanye 2023). The COVID-19 pandemic affected cities across Zimbabwe, with some areas experiencing higher infection and transmission rates than others (Chirisa et al. 2021). The pandemic affected urban areas more due to their higher population densities and limited access to healthcare and other essential services. However, the pandemic negatively impacted cities of all sizes across the country (Sharifi and Khavarian-Garmsir 2020). Table 4.1 shows confirmed COVID-19 cases in Zimbabwe by province as of June 17, 2022.

Regarding distribution, Harare was the hardest hit, with the highest number of confirmed cases and deaths. Other cities such as Bulawayo and Chitungwiza reported significant cases (Makombe 2021); however, COVID-19 cases were reported in rural areas and smaller towns nationwide. Figure 4.6 shows the map of Zimbabwe with cumulative cases of COVID-19 in Zimbabwe by province, as of January 30, 2022. Harare had the highest cumulative cases at 46,255.

---

[2] The term "infodemic" refers to an overabundance of information, including both accurate and inaccurate information, about a particular topic, which makes it difficult for people to find trustworthy and reliable information. The COVID-19 pandemic was characterized by a significant infodemic, with an overwhelming amount of information circulating online and through social media. This created challenges for public health officials and individuals alike. Misinformation and conspiracy theories about the virus and the vaccines spread quickly, leading to confusion and mistrust in public health measures. This made it more difficult for public health officials to communicate accurate and important information about the pandemic and the measures needed to control its spread.

Table 4.1 Confirmed COVID-19 cases in Zimbabwe by province as of June 17, 2022

Province	Tests	Cumulative cases	Recovered	Active cases	Deaths
Mashonaland East	124	33,320 (+15)	32,567 (+6)	337	416 (0)
Harare	924	47,943 (+26)	45,815 (+58)	331	1,797 (+1)
Midlands	63	17,995 (+18)	17,296 (0)	270	429 (0)
Mashonaland West	0	31,308 (+8)	30,558 (+5)	170	580 (0)
Mashonaland Central	96	14,385 (+4)	13,916 (+3)	145	324 (0)
Manicaland	139	29,722 (+8)	29,029 (+20)	137	556 (0)
Matabeleland North	6	18,317 (+12)	18,068 (+1)	118	131 (0)
Bulawayo	582	22,093 (+9)	21,093 (+20)	117	883 (+2)
Masvingo	0	21,857 (+1)	21,535 (+10)	114	208 (0)
Matabeleland South	143	17,813 (+8)	17,502 (+7)	102	209 (+1)
Total	2,077	254,753 (+109)	247,379 (+120)	1,841	5,533 (+4)

Source: Ministry of Health and Child Care Zimbabwe (2022)

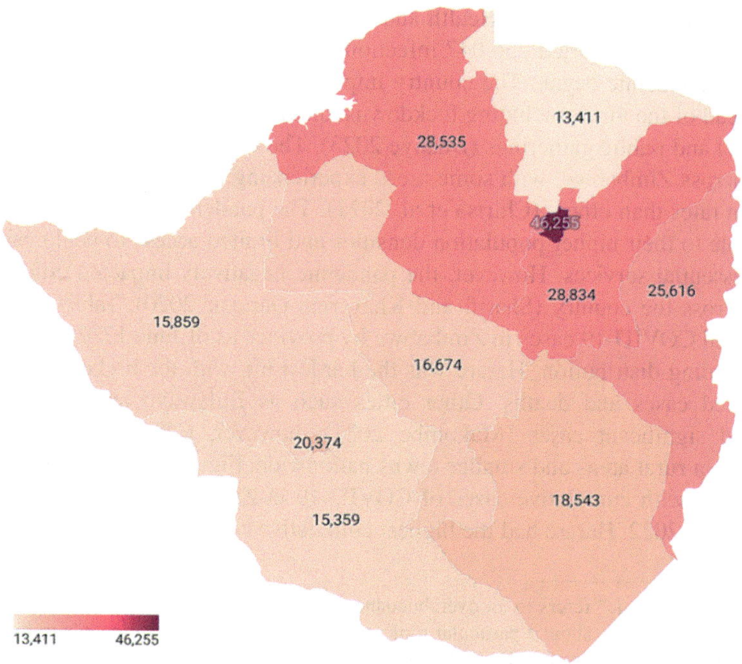

Fig. 4.6 Cumulative cases of COVID-19 in Zimbabwe by province, as of January 30, 2022. (Source: Ministry of Health and Childcare Zimbabwe 2022)

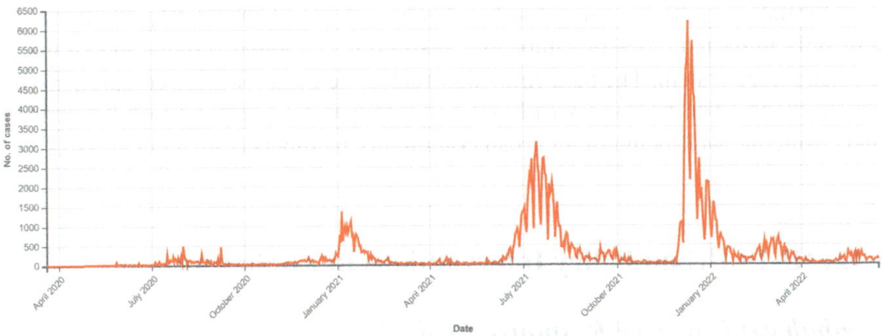

Fig. 4.7 Confirmed COVID-19 cases per day in Zimbabwe. (Source: Ministry of Health and Childcare Zimbabwe 2022)

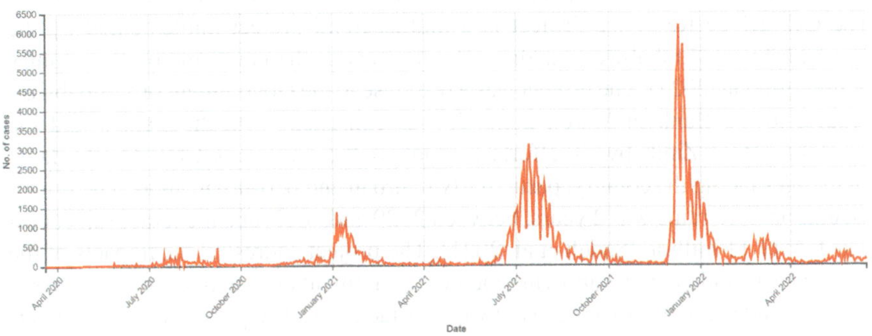

Fig. 4.8 Confirmed deaths per day in Zimbabwe. (Source: Ministry of Health and Childcare Zimbabwe 2022)

The severity of the COVID-19 pandemic varied from 2020 to 2022 with most cases recorded during the winter and rainy months, as graphically illustrated by Figs. 4.7 and 4.8. Figure 4.7 shows confirmed COVID-19 cases per day in Zimbabwe from April 2020 to April 2022. Figure 4.8 shows confirmed COVID-19 deaths per day in Zimbabwe from April 2020 to April 2022.

To mitigate the impact of the pandemic on cities, the government implemented a range of policy interventions, including increased access to healthcare and personal protective equipment, as well as financial support to businesses and individuals affected by the pandemic (OECD 2021). These measures reduced the virus' spread and mitigated effects on the population and economy. Local authorities in cities across Zimbabwe implemented measures such as the closure of nonessential businesses, the enforcement of social distancing and mask-wearing requirements, and the support of businesses and individuals affected by the pandemic (Mackworth-Young et al. 2021). It is worth noting that the COVID-19 pandemic significantly impacted the economy of Zimbabwe as a whole, with many businesses and individuals experiencing financial hardship due to the lockdowns and restrictions on activity (OECD 2020a, b, c, d). This negatively impacted the urban poor, who rely on informal economic activities for their livelihoods. The COVID-19 pandemic significantly impacted cities in Zimbabwe, disrupting economic activity and strained the healthcare system. The pandemic highlighted the need for more comprehensive and resilient urban policies, which are better prepared to respond to future crises and ensure the well-being of all urban community members (Chirisa et al. 2021).

COVID-19 Urban Policy Interventions in Zimbabwe

Zimbabwe: Covid-19 Response Project

Zimbabwe faced additional vulnerability challenges caused by the COVID-19 pandemic, which affected practically every country worldwide (Chirisa et al. 2021). The COVID-19 Response Project is a multisectoral project implemented by the Government of Zimbabwe to address the impacts of the pandemic on the country (Chitungo et al. 2022). The project was a response to the Government of Zimbabwe's request to the Bank in March 2020. It was consistent with the Bank's and the Government of Zimbabwe's efforts to address the enormous COVID-19-related challenges facing the country, particularly in urban and peri-urban areas. The implementation timeframe was 2 years, between 2020 and 2022, and the total project cost was UA 11.11 million (Sadeh et al. 2021). The project addressed the impacts of the COVID-19 pandemic by supplying medical equipment and supplies, training frontline responders, and providing handwashing facilities. The project provided emergency financial assistance to vulnerable households, supported small and medium enterprises affected by the pandemic, and improved the healthcare system's capacity to respond to COVID-19 (OECD 2020a, b, c, d).

The COVID-19 Response Project was implemented in two phases. The first phase was implemented from April 2020 to December 2020 and focused on emergency response. The second phase was implemented from 2021 to 2023 and focused on recovery and resilience (OECD 2020a, b, c, d). The project contributed in the limitation of COVID-19 morbidity and mortality rates by strengthening the response of the healthcare system to the pandemic. The direct beneficiaries were (i) the populations in high-density urban suburbs in Harare and satellite townships for improved access to handwashing facilities, (ii) 4500 community and village healthcare workers who were trained on COVID-19 prevention and case management protocols, and (iii) doctors and nurses in healthcare facilities who benefited from personal protective equipment and reduced infection rates. The indirect beneficiaries included the 680,000 population in the targeted suburbs and townships. The COVID-19 Response Project was implemented through a range of interventions, including the following:

(a) *Emergency cash transfers to vulnerable households to cover basic needs such as food and shelter. However, it is important to note that the distribution of government aid during a crisis can be complex and may not always reach the intended beneficiaries as planned when there is a misuse of funds.*

(b) *Support small to medium enterprises through the provision of grants and technical assistance.*

(c) *Investment in the healthcare system through the procurement of personal protective equipment and the expansion of testing and treatment capacity.*

(d) *Investment in the education system through the provision of distance learning materials and the expansion of digital infrastructure.*

(e) *Supporting the recovery of the economy by providing technical assistance to help businesses adapt to the new economic realities of the pandemic and promote economic growth.*

In addition, the COVID-19 Response Project supported the government's efforts to improve the country's preparedness for future health emergencies and to strengthen the resilience of communities and the economy to shocks and crises (OECD 2020a, b, c, d). Overall, the COVID-19 Response Project was an essential component of the government's efforts to address the impacts of the COVID-19 pandemic in Zimbabwe and to build a more resilient and prepared society.

Discussion of Policy Interventions on the Urban Poor

Prevention measures against crises like the COVID-19 pandemic must be appropriate in a local context. In Zimbabwe, communities require support with basic needs and access to reliable information to follow prevention measures (Chitungo et al. 2022). Healthcare workers urgently needed personal protective equipment and adequate salaries. Essential healthcare services and medications for conditions other than COVID-19 must also continue to be provided to reduce excess mortality and morbidity (Jessop et al. 2020).

The COVID-19 pandemic disproportionately impacted poor urban communities, as these communities lacked access to resources and support systems to cope with the economic and social impacts of the pandemic (Pachauri and Zimm 2021). In Zimbabwe, the government implemented policy interventions to address the needs of the urban poor during the pandemic. One such intervention was cash transfers and support for the informal sector. Many poor urban communities in Zimbabwe rely on informal economic activities, such as street vending and small-scale trade for their livelihoods (Chirisa et al. 2021). The lockdowns and restrictions on public gatherings in response to the pandemic significantly impacted these activities, which led to a loss of income for many poor urbanites. The government's cash transfer program provided financial support to these individuals to mitigate the pandemic's economic impacts; however, the cushioning allowances were hardly enough for the urban poor (GIGA 2020).

Another policy intervention was the expansion of healthcare services in urban areas. The COVID-19 pandemic strained healthcare systems worldwide, particularly in urban areas where the concentration of people made it more challenging to contain the spread of the virus (De Groot and Lemanski 2021). The government's efforts to expand healthcare services in urban areas, including the provision of additional beds and the expansion of testing and treatment capacities, improved access to healthcare for the urban poor and reduced the burden on the healthcare system (Tessema et al. 2021).

In response to the COVID-19 pandemic, the government and local authorities in Zimbabwe implemented policy interventions to support the urban poor and mitigate the impacts of the pandemic. These interventions included the following:

(a) *Financial assistance: the government provided cash transfers and other financial assistance to households and businesses affected by the pandemic. This included support for the informal sector, which is often a key source of livelihood for the urban poor.*
(b) *Expanding access to healthcare: the government implemented measures to increase access to healthcare services for the urban poor, including expanding testing and treatment capacities and providing personal protective equipment to healthcare workers.*
(c) *Supporting the informal sector: the government implemented initiatives to support the informal sector, including providing technical assistance and other forms of support to help businesses adapt to the new economic realities of the pandemic.*

To some extent, the policy interventions mitigated the impacts of the COVID-19 pandemic on the urban poor in Zimbabwe (World Bank 2021a, b). While these policies were designed to protect public health and mitigate the economic impact of the pandemic, they had unintended consequences for the urban poor. For example, lockdowns and social distancing measures disproportionately impacted the urban poor, who lived in crowded and cramped conditions. In many cases, these measures made it difficult or impossible for them to practice social distancing, increasing their risk of contracting and spreading the virus. Additionally, the closure of nonessential businesses and services severely impacted the livelihoods of the urban poor, many of whom relied on daily wages and informal work. Many governments implemented financial aid programs to mitigate the economic impact of the pandemic. However, these programs were often inaccessible or insufficient for the urban poor.

Policy discrimination against the urban poor is not a new phenomenon in Zimbabwe and Africa in general; it a longstanding issue that the COVID-19 pandemic has exacerbated. The pandemic highlighted the unequal distribution of resources and opportunities between different groups, with the urban poor particularly vulnerable. Discrimination against the urban poor took many forms in Zimbabwe, including a lack of access to healthcare, education, and basic services. To further demonstrate the urban penalty of the COVID-19 pandemic policies, the implementation of lockdowns and social distancing measures was accompanied by increased policing and harassment of the urban poor, who were perceived as violating the rules more frequently than other groups. Additionally, financial aid programs have often been designed to exclude the urban poor, perpetuating the existing discrimination against this group.

The unfair, discriminatory, and unjust treatment of the urban poor through COVID-19 policy was not only limited to Zimbabwe. Case studies from around the world also illustrate the impact of COVID-19 policy interventions on the urban poor and the perpetuation of discrimination. In India, for example, the nationwide lockdown imposed in March 2020 resulted in the displacement of millions of migrant workers, many of whom were left stranded without food or shelter. The lockdown resulted in the closure of informal businesses and services, affecting the livelihoods of the urban poor. In other African countries like South Africa, the series of nationwide lockdowns by the government resulted in the displacement of thousands of people living in informal settlements. The government's response to the pandemic was criticized for being insensitive to the needs of the urban poor, who were left without access to basic services such as water and sanitation.

From the discussion, it is essential to note that more could have been done and still needs to be done in terms of policy to address the underlying structural issues, which contribute to the vulnerability and continued marginalization of urban communities in Zimbabwe, Africa, and beyond. This includes addressing issues such as inadequate housing, lack of clean water and sanitation, and limited access to education and healthcare.

Conclusion

This chapter examined COVID-19 and urban policy interventions in Zimbabwe. The chapter presented the different approaches that can be taken when developing urban policies in times of crisis, the rational comprehensive policy, incremental planning policies, and strategic planning policies, each with strengths and weaknesses. The chapter presented the different ideologies or philosophical beliefs about the role and function of government, the economy, and society influencing policy formulation. These range from socialist, neoliberal, capitalist, people-centered, bottom-up/top-down, and interventionist policies. During the COVID-19 pandemic, the majority of policy interventions were top-down. The chapter highlighted the rationale for COVID-19 policy interventions in Zimbabwe, focusing on the WHO role in the policy terrain, SADC policy perspectives on COVID-19, and COVID-19 urban policy interventions in Zimbabwe. The COVID-19 pandemic significantly impacted urban policy in Zimbabwe, as the government had to implement measures rapidly to contain the spread of the virus and mitigate its effects on the population and economy (Mutanda 2022).

Interventions included lockdowns, quarantine measures, travel and public gathering restrictions, initiatives to provide financial support to businesses and individuals affected by the pandemic, increase access to healthcare, and personal protective equipment for healthcare workers. To address the needs of the urban poor, the government implemented measures such as financial assistance, health interventions, and housing and shelter. To some extent, policy interventions in response to the COVID-19 pandemic addressed the urban poor's needs and mitigated the pandemic's impact on the urban poor. However, as the chapter revealed, most of the policies were top-down and interventionistic as they were directives from the WHO, which dictated global interventions, and the national governments. Subsequently, these policies had little regard for the welfare of the urban poor who struggled to make ends meet and worsened their suffering.

In conclusion, the COVID-19 pandemic exposed and exacerbated inequalities and discrimination against the urban poor. While the Zimbabwean government and governments worldwide implemented policy interventions to mitigate the impact of the pandemic, these policies were often ineffective or even harmful to the urban poor. Lockdowns and social distancing measures disproportionately affected the living conditions and livelihoods of the urban poor, while financial aid programs were inaccessible or insufficient. Discrimination against the urban poor was perpetuated and exacerbated by the pandemic, with increased policing and harassment and exclusion from financial aid programs. Cases around the globe also demonstrate the penalization of the urban poor through COVID-19 pandemic policies and the perpetuation of discrimination.

To address failures, governments must ensure that policy interventions during crises like the COVID-19 pandemic are inclusive and accessible to the urban poor. In this way, the underlying social and economic inequalities perpetuating discrimination are addressed. This includes improved access to healthcare, education, and basic services and support of informal workers and businesses. By addressing the needs of the urban poor, governments can build more resilient and equitable societies, better prepared to face future crises. Further, the COVID-19 pandemic highlighted the importance of comprehensive and resilient urban policies better prepared to respond to future crises. The government and local authorities need to continue monitoring and addressing the needs of the urban poor in the long-term to ensure that they fully participate in and benefit from recovery processes. In the future, it is essential to consider the lessons learned from the COVID-19 pandemic and develop more comprehensive and resilient urban policies to respond to future crises. This will require the collaboration of government, civil society, and private sector as well as a commitment to address the needs of the most vulnerable and marginalized members of society.

References

Anwar A, Malik M, Raees V, Anwar A (2020) Role of mass media and public health communications in the COVID-19 pandemic. Cureus 12(9):e10453. https://doi.org/10.7759/cureus.10453

Ayouni I, Maatoug J, Dhouib W, Zammit N, Fredj SB, Ghammam R, Ghannem H (2021) Effective public health measures to mitigate the spread of COVID-19: a systematic review. BMC Public Health 21(1):1015. https://doi.org/10.1186/s12889-021-11111-1

Barnes MD, Hanson CL, Novilla LM, Meacham AT, McIntyre E, Erickson BC (2008) Analysis of media agenda setting during and after Hurricane Katrina: implications for emergency preparedness, disaster response, and disaster policy. Am J Public Health 98(4):604–610. https://doi.org/10.2105/AJPH.2007.112235

Bennett C, Howlett M (1992) The lessons of learning: reconciling theories of policy learning and policy change. Policy Sci 25:275–294

Bhandary RR, Gallagher KS, Zhang F (2021) Climate finance policy in practice: a review of the evidence. Clim Pol 21(4):529–545. https://doi.org/10.1080/14693062.2020.1871313

Bhanye J (2023) “Nimble sociality and belonging”: an ethnography of migrants’ responses to bans on associational life during the COVID-19 pandemic. In: Urban forum. Springer Netherlands, Dordrecht, pp 1–27. https://doi.org/10.1007/s12132-023-09503-0

Bhanye J, Bhanye A (2023) A blessing or a curse?: the role of social media during the COVID-19 pandemic in Africa. In: The Palgrave handbook of global social problems. Palgrave Macmillan, Cham. https://doi.org/10.1007/978-3-030-68127-2_367-1

Bowles P (2007) Capitalism. Pearson/Longman, New York

Buseh AG, Stevens PE, Bromberg M, Kelber ST (2015) The Ebola epidemic in West Africa: challenges, opportunities, and policy priority areas. Nurs Outlook 63(1):30–40

Cairney P, Oliver K (2017) Evidence-based policy-making is not like evidence-based medicine, so how far should you go to bridge the divide between evidence and policy? Health Res Policy Syst 15:35. https://doi.org/10.1186/s12961-017-0192-x

Carlquist E, Phelps J (2014) Neoliberalism. In: Teo T (ed) Encyclopedia of critical psychology. Springer, New York. https://doi.org/10.1007/978-1-4614-5583-7_390

Cerna L (2014) The nature of policy change and implementation: a review of different theoretical approaches. Organisation for Economic Cooperation and Development (OECD) report. 26:492–502

Charrua AB, Padmanaban R, Cabral P, Bandeira S, Romeiras MM (2021) Impacts of the tropical cyclone Idai in Mozambique: a multi-temporal landsat satellite imagery analysis. Remote Sens 13(2):201. https://doi.org/10.3390/rs13020201

Chirisa I, Mavhima B, Nyevera T, Chigudu A, Makochekanwa A, Matai J, Masunda T, Chandaengerwa EK, Machingura F, Moyo S, Chirisa H, Mhloyi M, Murwira A, Mhandara L, Katsande R, Muchena K, Manjeya E, Nyika T, Mundau L (2021) The impact and implications of COVID-19: reflections on the Zimbabwean society. Soc Sci Humanit Open 4(1):100183. https://doi.org/10.1016/j.ssaho.2021.100183

Chirisa I, Mutambisi T, Chivenge M (2022) The urban penalty of COVID-19 lockdowns across the globe: manifestations and lessons for Anglophone sub-Saharan Africa. GeoJournal 87:815–828. https://doi.org/10.1007/s10708-020-10281-6

Chitungo I, Dzinamarira T, Tungwarara N, Chimene M, Mukwenha S, Kunonga E, Musuka G, Murewanhema G (2022) COVID-19 response in Zimbabwe: the need for a paradigm shift? COVID 2(7):895–906. https://doi.org/10.3390/covid2070065

Cingano F (2014) Trends in income inequality and its impact on economic growth, OECD social, employment and migration working papers, no. 163. OECD Publishing. https://doi.org/10.1787/5jxrjncwxv6j-en

Cook S, Yi I (2011) Combating poverty and inequality through social policies: reflections on the UNRISD report. Glob Soc Policy. 11(2-3):135–137

Cuneo CN, Sollom R, Beyrer C (2017) The cholera epidemic in Zimbabwe, 2008–2009: a review and critique of the evidence. Health Hum Rights 19(2):249–264

De Groot J, Lemanski C (2021) COVID-19 responses: infrastructure inequality and privileged capacity to transform everyday life in South Africa. Environ Urban 33(1):255–272. https://doi.org/10.1177/0956247820970094

de Vreese CH, Esser F, Aalberg T, Reinemann C, Stanyer J (2018) Populism as an expression of political communication content and style: a new perspective. Int J Press Polit 23(4):423–438. https://doi.org/10.1177/1940161218790035

Dzobo M, Chitungo I, Dzinamarira T (2020) COVID-19: a perspective for lifting lockdown in Zimbabwe. Pan Afr Med J 35(Suppl 2):13. https://doi.org/10.11604/pamj.2020.35.2.23059

Fisher KA, Tenforde MW, Feldstein LR, Lindsell CJ, Shapiro NI, Files DC, Gibbs KW, Erickson HL, Prekker ME, Steingrub JS, Exline MC (2020) Community and close contact exposures associated with COVID-19 among symptomatic adults≥ 18 years in 11 outpatient health care facilities—United States, July 2020. Morb Mortal Wkly Rep 69(36):1258

Forester J (1984) Bounded rationality and the politics of muddling through. Public Adm Rev 44(1):23–31. https://doi.org/10.2307/975658

Gatwiri K, Amboko J, Okolla D (2020) The implications of neoliberalism on African economies, health outcomes and well-being: a conceptual argument. Soc Theory Health 18(1):86–101. https://doi.org/10.1057/s41285-019-00111-2

GIGA Focus Africa (2020) Africa after the Covid-19 lockdowns: economic impacts and prospects. Available at: https://www.giga-hamburg.de/en/publications/giga-focus/africa-after-the-covid-19-lockdowns-economic-impacts-and-prospects

Gilmore B, Ndejjo R, Tchetchia A (2020) Community engagement for COVID-19 prevention and control: a rapid evidence synthesis. BMJ Glob Health 5:e003188

Gkiotsalitis K, Cats O (2021) Public transport planning adaption under the COVID-19 pandemic crisis: literature review of research needs and directions. Transp Rev 41(3):374–392. https://doi.org/10.1080/01441647.2020.1857886

Güner R, Hasanoğlu I, Aktaş F (2020) COVID-19: prevention and control measures in community. Turk J Med Sci 50(SI-1):571–577. https://doi.org/10.3906/sag-2004-146

Guo C, Hu X, Yuan D (2022) The effect of COVID-19 on public confidence in the World Health Organization: a natural experiment among 40 countries. Glob Health 18:77. https://doi.org/10.1186/s12992-022-00872-

Haider N, Osman AY, Gadzekpo A (2020) Lockdown measures in response to COVID-19 in nine sub-Saharan African countries. BMJ Glob Health 5:e003319

Haque MS (1999) The fate of sustainable development under neo-liberal regimes in developing countries. Int Polit Sci Rev/Rev Int Sci Polit 20(2):197–218. http://www.jstor.org/stable/1601575

Harrison K, Boyd T (2018) The role of ideology in politics and society. In: Understanding political ideas and movements. Manchester University Press, Manchester. Available from: https://doi.org/10.7765/9781526137951.00011. Accessed 27 Dec 2022

Harrison CG, Williams PR (2016) A systems approach to natural disaster resilience. Simul Model Pract Theory 65:11–31. https://doi.org/10.1016/j.simpat.2016.02.008

Harvey D (2007) Neoliberalism as creative destruction. Ann Am Acad Pol Soc Sci 610:22–44. http://www.jstor.org/stable/25097888

Hofer E, Musakwa W, van Lanen S (2022) Inclusivity insights: two urban development projects in Johannesburg. J House Built Environ 37:1835–1858. https://doi.org/10.1007/s10901-021-09916-y

Horn P, Mitlin D, Bennett J, Chitekwe-Biti B, Makau J (2018) Towards citywide participatory planning: emerging community-led practices in three African cities. Global Development Institute, Working Paper Series 1(34)

Hudson B, Hunter D, Peckham S (2019) Policy failure and the policy-implementation gap: can policy support programs help? Policy Design Pract 2(1):1–14. https://doi.org/10.1080/25741292.2018.1540378

Hyland-Wood B, Gardner J, Leask J (2021) Toward effective government communication strategies in the era of COVID-19. Humanit Soc Sci Commun 8(30). https://doi.org/10.1057/s41599-020-00701-w

Jessop ZM, Dobbs TD, Ali SR, Combellack E, Clancy R, Ibrahim N, Jovic TH, Kaur AJ, Nijran A, O'Neill TB, Whitaker IS (2020) Personal protective equipment for surgeons during COVID-19 pandemic: systematic review of availability, usage and rationing. Br J Surg 107(10):1262–1280. https://doi.org/10.1002/bjs.11750

Kollamparambil U, Oyenubi A (2021) Behavioural response to the Covid-19 pandemic in South Africa. PLoS One 16(4):e0250269. https://doi.org/10.1371/journal.pone.0250269

Koontz TM, Newig J (2014) From planning to implementation: top-down and bottom-up approaches for collaborative watershed management. Policy Stud J 42:416–442

Kørnøv L, Thissen WA (2000) Rationality in decision-and policy-making: implications for strategic environmental assessment. Impact Assess Project Appraisal 18(3):191–200

Krause NM, Freiling I, Beets B, Brossard D (2020) Fact-checking as risk communication: the multi-layered risk of misinformation in times of COVID-19. J Risk Res 23(7–8):1052–1059

Kuruvilla S, Dorstewitz P (2010) There is no "point" in decision-making: a model of transactive rationality for public policy and administration. Policy Sci 43:263–287

Lapuente V, Van de Walle S (2020) The effects of new public management on the quality of public services. Governance 33(3):461–475

Lindblom CE (1959) The science of muddling through. Public Adm Rev 19:79–88

Luzon JL (1988) Housing in socialist Cuba: an analysis using Cuban censuses of population and housing. Cuba Stud 18:65–83. http://www.jstor.org/stable/24486955

Mackworth-Young CR, Chingono R, Mavodza C, McHugh G, Tembo M, Chikwari CD, Weiss HA, Rusakaniko S, Ruzario S, Bernays S, Ferrand RA (2021) Community perspectives on the COVID-19 response, Zimbabwe. Bull World Health Organ 99(2):85–91. https://doi.org/10.2471/BLT.20.260224

Makombe EK (2021) "Between a rock and a hard place": the coronavirus, livelihoods, and socio-economic upheaval in Harare's high-density areas of Zimbabwe. J Dev Soc 37(3):275–301. https://doi.org/10.1177/0169796X211030062

Masiya T, Mandiyanike D, Molosiwa D, Mazenda A (2021) Southern African responses to the COVID-19 pandemic: a study of Botswana and South Africa. Afr Public Serv Deliv Perform Rev 9(1):a497. https://doi.org/10.4102/apsdpr.v9i1.497

Matamanda AR, Nel V, Leboto-Khetsi L, Dunn M (2022) Risk communication in an informal settlement during COVID-19: case of Dinaweng, Bloemfontein South Africa. Urban Governance. https://doi.org/10.1016/j.ugj.2022.10.002

Maulani N, Nyadera IN, Wandekha B (2020) The generals and the war against COVID-19: the case of Zimbabwe. J Glob Health 10(2):020388

Ministry of Health and Child Care [MoHCCZim] (2022) COVID-19 update: as at 17 June 2022, Zimbabwe had 254 753 confirmed cases, including 247 379 recoveries and 5 533 deaths. To date, a total of 6 260 228 people have been vaccinated against COVID-19, June 17

Murewanhema G, Burukai T, Mazingi D, Maunganidze F, Mufunda J, Munodawafa D, Pote W (2020) A descriptive study of the trends of COVID-19 in Zimbabwe from March–June 2020: policy and strategy implications. Pan Afr Med J 37(Suppl 1):33. https://doi.org/10.11604/pamj.supp.2020.37.33.25835

Murewanhema G, Musuka G, Denhere K, Chingombe I, Mapingure MP, Dzinamarira T (2022) The landscape of COVID-19 vaccination in Zimbabwe: a narrative review and analysis of the strengths, weaknesses, opportunities and threats of the programme. Vaccine 10(2):262. https://doi.org/10.3390/vaccines10020262

Mutanda D (2022) Challenges and opportunities for Zimbabwe's responses to COVID-19. Cogent Soc Sci 8(1). https://doi.org/10.1080/23311886.2022.2084890

Navarro V (2007) Neoliberalism as a class ideology; or, the political causes of the growth of inequalities. Int J Health Serv 37(1):47–62. http://www.jstor.org/stable/45138405

Nicola M, Alsafi Z, Sohrabi C, Kerwan A, Al-Jabir A, Iosifidis C, Agha M, Agha R (2020) The socio-economic implications of the coronavirus pandemic (COVID-19): a review. Int J Surg (London, England) 78:185–193. https://doi.org/10.1016/j.ijsu.2020.04.018

OECD (2017) Debate the issues: complexity and policy making, OECD insights. OECD Publishing, Paris. https://doi.org/10.1787/9789264271531-en

OECD (2020a) Building back better: a sustainable, resilient recovery after COVID-19, OECD policy responses to coronavirus (COVID-19). OECD Publishing, Paris. https://doi.org/10.1787/52b869f5-en

OECD (2020b) Government support and the COVID-19 pandemic, OECD policy responses to coronavirus (COVID-19). OECD Publishing, Paris. https://doi.org/10.1787/cb8ca170-en

OECD (2020c) Tax and fiscal policy in response to the coronavirus crisis: strengthening confidence and resilience, OECD policy responses to coronavirus (COVID-19). OECD Publishing, Paris. https://doi.org/10.1787/60f640a8-en

OECD (2020d) The impact of the coronavirus (COVID-19) crisis on development finance, OECD policy responses to coronavirus (COVID-19). OECD Publishing, Paris. https://doi.org/10.1787/9de00b3b-en

OECD (2021) The territorial impact of COVID-19: managing the crisis and recovery across levels of government, OECD policy responses to coronavirus (COVID-19). OECD Publishing, Paris. https://doi.org/10.1787/a2c6abaf-en

OCHA (2021) Global humanitarian overview 2021. https://2021.gho.unocha.org/. Accessed 20 Jan 2022

Orlove B, Shwom R, Markowitz E, Cheong S (2020) "Climate decision-making." Annu Rev Environ Resour 45(1):271–303

Pachauri S, Zimm C (2021) Deprivations and inequities in cities viewed through a pandemic lens. Front Sustain Cities 3. https://doi.org/10.3389/frsc.2021.645914

Park MB, Ranabhat CL (2022) COVID-19 trends, public restrictions policies and vaccination status by economic ranking of countries: a longitudinal study from 110 countries. Arch Public Health 80:197. https://doi.org/10.1186/s13690-022-00936-w

Roose A, Kull A (2012) Empowering spatial information in the evolution of planning systems: lessons of ad-hoc plans in Estonia. Reg Stud 46(4):493–508. https://doi.org/10.1080/00343404.2011.653336

Rosenberg J, Strauss I, Isaacs G (2021) COVID-19 impact on SADC labour markets: evidence from high-frequency data and other sources. Afr Dev Rev/Rev Afr Dev 33(Suppl 1):S177–S193. https://doi.org/10.1111/1467-8268.12528

Rwodzi A, Fosl PS (2020) Reconciliation: a false start in Zimbabwe? (1980–1990). Cogent Arts Humanit 7(1). https://doi.org/10.1080/23311983.2020.1745555

Rydin Y, Bleahu A, Davies M, Dávila JD, Friel S, De Grandis G, Groce N, Hallal PC, Hamilton I, Howden-Chapman P, Lai KM, Lim CJ, Martins J, Osrin D, Ridley I, Scott I, Taylor M, Wilkinson P, Wilson J (2012) Shaping cities for health: complexity and the planning of urban environments in the 21st century. Lancet (London, England) 379(9831):2079–2108. https://doi.org/10.1016/S0140-6736(12)60435-8

Sadeh A, Radu CF, Feniser C, Borsa A (2021) Governmental intervention and its impact on growth, economic development, and technology in OECD countries. Sustainability 13(166). https://doi.org/10.3390/su13010166

Santana MJ, Manalili K, Jolley RJ, Zelinsky S, Quan H, Lu M (2018) How to practice person-centred care: a conceptual framework. Health Expect 21(2):429–440. https://doi.org/10.1111/hex.12640

Scheba A, Turok I, Visagie J (2021) The role of social housing in reducing inequality in South African cities. AFD Research Papers, pp 1–79. https://www.cairn-int.info/journal%2D%2D2021-202-page-1.htm

Sharifi A, Khavarian-Garmsir AR (2020) The COVID-19 pandemic: impacts on cities and major lessons for urban planning, design, and management. Sci Total Environ 749:142391. https://doi.org/10.1016/j.scitotenv.2020.142391

Steinmetz J (2019) Ideologies of the individual in politics, power, and purpose: an orientation to political science. Available at: https://fhsu.pressbooks.pub/orientationpolisci/chapter/chapter-2/. Accessed 20 Jan 2022

Tessema GA, Kinfu Y, Dachew BA (2021) The COVID-19 pandemic and healthcare systems in Africa: a scoping review of preparedness, impact and response. BMJ Glob Health 6:e007179

The Herald Zimbabwe (2021) Covid-19 case no 1 speaks. https://www.herald.co.zw/covid-19-case-no-1-speaks/. The Herald, 9 April 2021. Accessed 19 Jan 2023

Turner MJ (2016) Rational emotive behavior therapy (REBT), irrational and rational beliefs, and the mental health of athletes. Front Psychol 7:1423. https://doi.org/10.3389/fpsyg.2016.01423

UN-HABITAT (2020) Participatory incremental urban planning. A toolbox to support local governments in developing countries to implement the new urban agenda and the sustainable development goals. United Nations Human Settlements Programme, Nairobi

Uzonwanne FC (2016) Rational model of decision making. In: Farazmand A (ed) Global encyclopedia of public administration, public policy, and governance. Springer, Cham. https://doi.org/10.1007/978-3-319-31816-5_2474-1

Van der Waldt G (2015) Government interventionism and sustainable development: the case of South Africa. Afr J Public Aff 8(2):35–50

Weible CM, Nohrstedt D, Cairney P (2020) COVID-19 and the policy sciences: initial reactions and perspectives. Policy Sci 53:225–241. https://doi.org/10.1007/s11077-020-09381-4

WHO (2020) Coronavirus disease (COVID-19) pandemic. https://www.who.int/europe/emergencies/situations/covid-19. Accessed 20 Jan 2022

World Bank (2021a) Governance & institutions COVID-19 response resources. Available at: https://www.worldbank.org/en/topic/governance/brief/governance-institutions-covid-19-response-resources. Accessed 20 Jan 2022

World Bank (2021b) Zimbabwe economic update: COVID-19 further complicates Zimbabwe's economic and social conditions. Available at: https://www.worldbank.org/en/country/zimbabwe/publication/zimbabwe-economic-update-covid-19-further-complicates-zimbabwe-s-economic-and-social-condition. Accessed 20 Jan 2022

World Health Organization (2017) A strategic framework for emergency preparedness. Available at: https://www.preventionweb.net/publication/strategic-framework-emergency-preparedness

World Health Organization (2021) Rapid tests up Zimbabwe's COVID-19 diagnosis. Available at: https://www.afro.who.int/news/rapid-tests-zimbabwes-covid-19-diagnosis. Accessed 20 Jan 2022

York G (2021) Mounting desperation in Africa as COVID-19 vaccine shortage persists. The Globe and Mail, Harare and Johannesburg

Chapter 5
The COVID-19 Lockdowns and Poor Urbanites in Harare, Zimbabwe: Exploring Socioeconomic Impacts with Remote Ethnography

Introduction

An estimated 90% of all reported COVID-19 cases made urban areas the epicenters of the COVID-19 pandemic (UN-Habitat 2020a, b). Cities were particularly vulnerable to the spread of the virus because of the population size and high level of global and local interconnectivity. For many African cities, the COVID-19 health crisis expanded to a crisis of urban access, urban equity, urban finance, safety, joblessness, public services, infrastructure, and transport, all disproportionally affecting the most vulnerable urbanites (Lone and Ahmad 2020; UN-Habitat 2020b). Tackling the COVID-19 pandemic was also challenging in African urban areas because of high levels of crime and violence, poor infrastructure and housing, weak local governance, and ill-equipped or under-resourced frontline workers (Bhanye 2023a; Bhardwaj et al. 2020). Among African city inhabitants, the COVID-19 crisis disproportionately affected informal settlement dwellers compared to other urban residents, particularly those in low- and middle-income countries (Matamanda et al. 2022; Mulugeta et al. 2021; Turok and Visagie 2021). Because of informal settlements' physical, structural, and social aspects (Bhanye 2022, 2023b), prevention efforts, such as handwashing, self-isolation, and physical distancing, were not feasible. The health impacts of COVID-19 in informal settlements were intensified by poor access to health services, often within those settings (Tampe 2021).

One of the most significant policy responses to the COVID-19 pandemic was mandatory lockdowns[1] and restrictions on movement to mitigate the spread of the

[1] COVID-19 lockdown measures refer to a set of restrictions and regulations put in place by governments and health authorities to slow the spread of COVID-19, a highly infectious respiratory illness caused by the SARS-CoV-2 virus. These measures limit social contact between individuals and reduce the opportunities for the virus to spread from person to person. Lockdown measures can vary depending on the severity of the outbreak and the response of the government in different

J. I. Bhanye et al., *COVID-19 Lockdowns and the Urban Poor in Harare, Zimbabwe*, https://doi.org/10.1007/978-3-031-41669-9_5

virus (Matsungo and Chopera 2020; Organization for Economic Cooperation and Development (OECD) 2021; Bhanye 2023). Like other governments across the globe, the Government of Zimbabwe implemented strict lockdown measures to control the spread of the virus. The Government of Zimbabwe declared a state of national disaster in response to the COVID-19 pandemic on Friday, March 27, 2020 (Herald 2020). This was followed by the first intense nationwide lockdown on March 30, 2020, resulting in intercity travel bans, closure of educational institutions, workplaces, stadiums, theatres, and shopping centers, and mandatory quarantines of potentially infected people (Herald 2020). From there, a series of mandatory lockdowns followed until 2022. Containment measures as a policy strategy for COVID-19 had its own benefits, including reduced hospital admissions and deaths from the virus. Indeed, some of the densest cities of the developed world managed to bring initial outbreaks of COVID-19 under control with a very low incidence of infections and deaths (Atalan 2020; Lau et al. 2020; OECD 2021). For example, Australia, Japan, and South Korea brought COVID-19 prevalence down dramatically—including cities like Seoul, Sydney, and Tokyo—emphasizing anticipation, early preparation, and a proactive approach when caseloads were still low and using mitigation measures such as mask wearing (OECD 2021). However, many studies point out the far-reaching negative impacts of lockdowns, particularly in developing countries; the pandemic placed a magnifying glass on existing inequalities and poverty in developing countries (Teachout and Zipfel 2020; UN-Habitat 2020a).

The economic impacts of the COVID-19 pandemic and mandatory lockdowns were profound. The early studies on the impacts of COVID-19 mandatory lockdowns reveal that the policy led to much larger and more rapid contractions of economic activity than seen in previous crises, including the global food crisis of 2007–2008 and the 2009 recession (Teachout and Zipfel 2020; Thurlow 2020). A study by the International Labor Organization (ILO 2020) estimated that the pandemic led to a loss of 25 million jobs worldwide. The lockdown measures significantly reduced economic activity, forcing many businesses to shut down. Globally, the effects were especially severe in the informal sector, which represented 90% and 67% of total employment in low- and middle-income countries, respectively (ILO 2020). In the first month of the crisis, informal workers worldwide lost an average of 60% of their earnings. In Africa and Latin America, this figure was nearly 80% (United Nations (UN) 2020). This has had devastating impacts on women. Globally, women are overrepresented in the informal economy and the hardest-hit sectors, such as tourism, hospitality, and services (UN-Habitat 2020b). In developing countries, many urban poor depend on daily incomes from informal work; lockdown policies undermined such households' survival opportunities (Avetisyan 2020; Chirisa et al. 2020; Matamanda et al. 2022).

The mandatory lockdowns caused significant social impacts (Adams-Prassl et al. 2020; Gisselquist and Kundu 2020). Deep-rooted inequalities, including where in a

---

countries, but some common examples include stay-at-home orders, closure of nonessential businesses, restrictions on travel, limits on public gatherings, and mandatory mask wearing (WHO 2020).

city a person lives and works and a person's gender and age, led to the pandemic having a disproportional impact on groups that were already vulnerable (Balde et al. 2020; Espi et al. 2020; Schotte et al. 2021). Lockdowns increased social isolation, with many people experiencing loneliness and mental health issues (Debuquet et al. 2020). The closure of schools significantly impacted children's education, with many experiencing a disruption to their learning (OECD 2020). In the poorest parts of the world, more people became poor and food insecure as their livelihoods depended on work within informal sectors, which was halted by mandatory lockdowns (Laborde et al. 2021; OECD 2021). Due to legal frameworks that do not officially recognize informal dwellers (Kimani-Murage et al. 2014), especially those of foreign origin, social welfare and economic stimulus packages did not reach those who needed these services most. Informal settlement dwellers that relied on their daily wages to feed themselves and their families frequently faced household food insecurity and were on the verge of malnutrition (Tampe 2021).

This chapter explores the socioeconomic impacts of COVID-19 mandatory lockdowns on poor urbanites in Harare, the capital city of Zimbabwe. Like many countries in Southern Africa, Zimbabwe has an under-resourced healthcare system, high unemployment, densely populated urban areas, and shortages of basic commodities (including water and food). This made mandatory lockdown policy measures difficult to adhere to and enforced among the urban poor (Mackworth-Young et al. 2021). A recent economic analysis study of the country demonstrates that the COVID-19 pandemic and mandatory lockdowns had devastating impacts on the urban poor, disrupting their livelihoods, expanding the number of impoverished citizens by 1.3 million, and increasing extreme poverty overall to 49% in 2020 (World Bank 2021). The report shows that nearly 500,000 Zimbabwean households had at least one member losing their job, causing many families to fall into poverty and worsening the plight of the poor. Food insecurity was exacerbated by inadequate reach/coverage of relevant social protection programs, particularly in poor, hard-to-reach[2] urban areas (World Bank 2021).

While there is a growing number of studies on the socioeconomic impacts of mandatory lockdowns on the urban poor in Zimbabwe, most of the studies used traditional research methodologies, for example, structured questionnaires (Matsungo and Chopera 2020), document analysis (Chirisa et al. 2020), and qualitative face to face interviews (Dzawanda et al. 2021; Nyanga and Zirima 2020). A desktop study by Makombe (2021) "Between a Rock and a Hard Place": The Coronavirus, Livelihoods, and Socioeconomic Upheaval in Harare's High-Density Areas of Zimbabwe highlights how the COVID-19 crisis led to widespread food insecurity, economic anxiety, and general disenfranchisement from alternative sources of income that in turn, created further social upheaval. Another study utilizing traditional qualitative methodologies and document analysis by Nyabeze and

[2] Hard-to-reach populations are groups of people who are difficult to engage, access, or reach through traditional methods of communication and outreach. These populations include people who are geographically isolated, are socially marginalized, or have limited access to healthcare, education, or other resources.

Chikoko (2021) on "Socio-economic impact of COVID-19 lockdown measures on the informal sector livelihoods in Zimbabwe" revealed the socioeconomic impact of COVID-19 on the informal sector such as disempowerment, exposure to poverty, and increased gender-based violence and inequality. However, traditional qualitative approaches like desktop research did not offer richer and more contextualized experiences of the urban poor during the COVID-19 pandemic. The COVID-19 pandemic presented serious challenges in doing face-to-face urban research, especially among poor, hard-to-reach populations. Yet, while the coronavirus has brought many challenges to fieldwork research, it also created opportunities for research to improve. The restrictions imposed during the COVID-19 pandemic sparked new interest in the practical and philosophical possibilities of "translating" traditional fieldwork into digital/remote format (Paupini et al. 2022). In our study, we innovatively embraced new research methodologies by adopting remote ethnography[3] to deepen our understanding of how the lockdown restrictions impacted the most vulnerable population in urban settings.

Since its emergence in the mid-nineteenth century, ethnography has had a strong tradition of physically researching communities. In classical ethnography, earliest ethnographers like Malinowski carried out research alone, in a place far from home and over a long-term period. However, ethnography continued to evolve, adapting to changes in the world system (Marcus 1995). A significant development of ethnography was when globalization and technology started to impact the second half of the twentieth century (Clifford and Marcus 1986; Kumoro 2022). Thus, digital ethnography is not a new phenomenon that arose with the COVID-19 pandemic (Paupini et al. 2022). Virtual ethnography came in the early twenty-first century due to the increasing new social order in the virtual world and the global Internet (Bhanye and Bhanye 2023; Bhanye et al. 2023; Kumoro 2022). However, some scholars warn that moving research online "risks reducing the complexity of social phenomena and omitting important aspects of lived experience" (Mwambari et al. 2021). They argue that online platforms and communication eclipse the immersion, context, and trust-building process with informants (Lawrence 2022).

There are also limits to virtual ethnographic practices, such as observing the community directly, participating in community activities directly, and, most importantly, experiencing everyday life practices from the community (Kumoro 2022). However, other scholars have explored the possibilities of transitioning research from physical to digital contexts in response to the COVID-19 pandemic. Lawrence and Moran and Caetano discuss the use of videoconferencing platforms such as Skype and Zoom for conducting qualitative interviews, and Watson and Lupton explore remote fieldwork in homes (Lawrence 2022; Moran and Caetano 2021; Watson and Lupton 2022). Since digital ethnography is often multi-sited,

[3] Digital ethnography (also known as "virtual ethnography," "cyber ethnography," "netnography," or "mobile ethnography") is a digital transformation of in-person ethnography and a form of online or remote ethnographic research that leverages the power of technology—gadgets and Internet—to help researchers generate rich remotely, contextual insights into the lived experiences of social groups under study (See Hjorth et al. 2017; Varis 2016).

researchers can pursue research across more than one connected site (Marcus 1995). Online interviews may also enable the recruitment of a wider range of participants regarding geographical location (as online interaction generally eliminates the need to travel) and the participants' availability (Bhanye and Bhanye 2023; Bhanye et al. 2023). Interviewees and interviewers tend to be more available for online communication since digital methods allow, for example, interviews to be scheduled in the evening, which might be impractical in person (Paupini et al. 2022). Paupini et al. (2022) highlight a new way of approaching fieldwork in homes by adopting "methodological improvisation" and proposes digital ethnography methods; ethnography is an option that enables data collection during a pandemic and is a sustainable alternative to traditional ethnography methods. Based on remote ethnographic empirical evidence, this chapter discusses socioeconomic impacts of COVID-19 lockdowns among the urban poor in Harare, Zimbabwe. The following section presents how we implemented digital ethnography in urban Harare.

Digital Ethnography During the COVID-19 Pandemic in Urban Harare

We adopted digital ethnographic methodologies after observing that COVID-19 posed significant difficulties to urban life and created challenges in conducting decent urban research because of mobility-related restrictions. With increasing access to mobile phone technology and Internet-mediated communications, we found digital ethnography as the most feasible methodology for our study. Using digital ethnography, we explored how COVID-19 and associated bans transformed everyday life of vulnerable urbanites across age, ethnicity, gender, religion, and cultural orientation. Participation in online spaces in Harare is increasing because of access to smartphones, 4G networks, and multimedia social sharing applications. We participated in online (covert-ethnography) and offline (overt-ethnography) digital ethnography. Additionally, leveraging mobile phones that have become an extension of everyday lives, especially among youth, we combined multiple digital platforms, including Twitter, Facebook, Instagram, WhatsApp, and YouTube, to explore the welfare of poor urbanites during the COVID-19 pandemic lockdowns.

Selection of Participants

To identify participants for social media platforms, especially WhatsApp, we utilized snowballing sampling technique; we systematically identified 50 urbanities, whom we incorporated into five WhatsApp focus group discussions and ten for COVID-19 diaries. The groups were categorized on participants' residential areas, economic activities, age groups, and social backgrounds. To select focus group participants, we began by identifying individuals in our residential areas who fit into categories of vulnerable urbanities or associated with vulnerable livelihood sources, such as people operating in informal sectors, living in informal settlements, and unemployed youths and women. After identifying the focal participants, we discussed with them verbally and through mobile phones the objectives of the study and how they could assist in identifying or inviting other participants with similar socioeconomic characteristics on a WhatsApp focus group discussion platform. A notable example was how we formed a pilot focus group discussion with youths who sold mobile phones and associated accessories. We first approached Stanely (aged 28 years), who survived on selling second-hand mobile phones. We explained in detail the study objectives, data protection, and confidentiality. We further asked him to invite 5–10 workmates to join a WhatsApp group he created for the discussion. After setting up a WhatsApp group, we introduced ourselves, the study's objectives and the parameters of participation during discussions. Like in our pilot focus group discussions, the other five groups were formed. These groups were based on economic activities—vendors, money changers, information and communication technology (ICT) dealers, taxi operators, and women in the informal sector.

Data Collection Procedures

During the study period, we were cognizant of how COVID-19 was trending in Zimbabwe. We followed related stories to search for new videos (skits) posted about COVID-19 on urban dynamics in Zimbabwe. To broaden our understanding of peoples' attitudes and coping strategies towards the lockdowns, we overtly took part in digital ethnography. We analyzed comments and discussions held on social media platforms such as blogs, Facebook, and Twitter for both individuals (particularly activists) and organizations. We followed short skits posted on YouTube, specifically by BustopTv,[4] ZimEye,[5] New ZimTV, and Open Parly Zimbabwe,[6] to gain a vivid picture of how society interprets lockdowns and associated challenges. For instance, we followed comments of a New ZimTV video on YouTube titled "We will die of hunger" Despair as Zimbabwe lockdown extended, posted three days

[4] Bustop Tv is a youth-run Zimbabwean media house established in 2014. It is known for satirical skits that go viral as they comment on prevailing socioeconomic and political issues affecting society. Bustop Tv has the most creative, talented, and experienced personnel in comedy, journalism, film, and video production. We listened to and followed comments for #BUSTOPTV #BUSTOP Lockdown 21 Days Series Episode.

[5] ZimEye is a media house, which specializes in providing intricate, detailed, and analytical news stories on the ground in Zimbabwe. It also covered issues on the plight of Zimbabweans during the COVID-19 crisis.

[6] Open Parly Zimbabwe is a leading online media platform covering parliament, politics, press conferences, and the everyday lives of Zimbabwe, focusing on the engagement between decision-makers and citizens for a better society for young Zimbabweans. Open Parly Zimbabwe is among the top media platforms covering COVID-19 issues in Zimbabwe.

after the first lockdown. We learned that such short clips and interviews posted on websites or media platforms reflected current dynamics on how the urban poor struggled to access basic needs (water, health, accommodation, and food) due to COVID-19 lockdowns. On online social media platforms such as Twitter Handle, we used phrases like #tagCOVID-19 in Harare, lockdown, shutdown, urban poor, etc., to search how participants reflected on their situations in the context of COVID-19. Internet-based platforms showed daily challenges encountered by thousands of poor urbanites during the COVID-19 pandemic, and diverse comments from subscribers of different backgrounds were obtained. Data from social media were collected manually through labelling social media posts, using, for example, Twitter hashtags like coronavirus, COVID-19, social distancing, and lockdown in Zimbabwe.

To deepen our understanding, we conducted 25 in-depth online interviews with key informants engaged in the informal sector, activists, church leaders, and civil society groups representing urbanities in Harare. The associations we engaged in included Epworth Residents Association, Harare Residents Alliance, and Survival Vendors Union of Zimbabwe. Having obtained a deep understanding of lockdowns, we conducted ground-verification assessments through observations in residential locations where some participants lived or did business. We conducted three transect walks in three different market locations (CBD, Mbare Market, and Kamunhu Shopping Centre), representing different socioeconomic consumer dynamics. The idea was to compare how socioeconomic background informed or shaped people's vulnerability and adaptation capacity to shocks and crises.

We gathered technical insights from four virtual meetings held by the Institute of Poverty and Agrarian Studies at Western Cape University to gather experts' views and appreciation of similar dynamics outside Zimbabwe. Notably, a webinar session we attended titled "The gendered impacts of COVID-19: Women in Informal work—03 August 2020" explored policy measures to engage women in the informal sector of the global south during adverse times like the COVID-19 pandemic. We reviewed secondary sources including newspapers, scientific journals, and reports from civic society groups and government agencies. A review of such literature deepened our understanding of the vulnerability of poor urbanities and extended insights into government interventions in such situations with resources or enabling policy frameworks. Much focus was on understanding gaps in existing social safety programs and on improving social welfare policy frameworks to accommodate concerns of already-marginalized populations during shocks. The literature review positioned us to have well-informed knowledge on the government's guidelines for social protection operations during shocks.

Ethical Considerations

We were conscious of ethical issues in all aspects of the research. The Ethical Committee of the University of the Free State granted ethical approval to conduct this study (UFS-HSD2020/1704/192). The Ethical Committee inspected our ethical guides at every stage of the research project: procedural ethics (how humans were protected before the research process), ethics in practice (how social welfare of defined social groups were protected), and ethics in publishing (how the authors protected participants' identities in reporting and publishing processes). Thus, key ethical issues included consent from all the participants before the research process. We designed an online consent form that the participants completed. The consent form informed participants on the following: (a) purpose, (b) methods to be used, (c) expected role of the participants, (d) intended use of findings, and (e) adherence to confidentiality.

Additionally, for social media platforms, like WhatsApp, participants were asked not to share media attachments or texts unrelated to the topic or objectives of the group. Just like when conducting face-to-face focus group discussions, we set a time and date for each group; each discussion lasted for about 30 min to 1 h and 15 min.

We discovered that in all focus group discussions participants preferred sending voice notes, videos, and photos to reflect on their everyday life in the context of the COVID-19 lockdown. To promote balanced participation during discussions, we randomly picked participants to share their ideas. To be precise, our digital ethnography was centered on reciprocal and mutual collaboration with participants. Using this approach, our urban research participants became co-participants and co-researchers rather than subjects of exploitation in the digital space. We emphasized human research ethical parameters guiding our study. More specifically, we articulated to participants that participation was voluntary, and as researchers we maintained confidentiality in all aspects to be shared/discussed by both participants and researchers.

Data Analysis

Collected data were thematically analyzed, where a theme was a patterned response or meaning that reflected answers to a research question. We began by familiarizing ourselves with the data we gathered; for instance, we listened to recorded discussions several times to understand the discussion and positionality of participants. We identified common texts that were prominent in all discussions and explicitly explored the meaning of the text and arranged (coded) it into themes. Texts with similar meanings were classified accordingly. Existing writings on ethical issues and digital ethnography were also used to shape the reporting process of our findings. In the following section, we present the findings from digital ethnography on the socioeconomic impacts associated with COVID-19 mandatory lockdowns among the urban poor in Harare.

Rights to the City, Urban Penalty, and Continued Marginalization of the Urban Poor: Evidence from Socioeconomic Impacts of COVID-19 Lockdowns

Through remote ethnography, this chapter provides evidence-based stories of how COVID-19 mobility restrictions affected poor urbanites in Harare, Zimbabwe. Our findings show four major socioeconomic impacts associated with COVID-19 mandatory lockdowns among the urban poor: food insecurity among poor urbanites, closure of the informal sector, induced urban–rural migration, and strengthening of visible urban disparities between the poor and the rich (Fig. 5.1).

We focus on these four major socioeconomic impacts of COVID-19 mandatory lockdowns as they are more relatable to aspects of marginalization, continued suffering, urban penalty, and threats to the rights to the city of the urban poor, which is the theoretical focus of this study. Regarding food insecurity among poor urbanites,

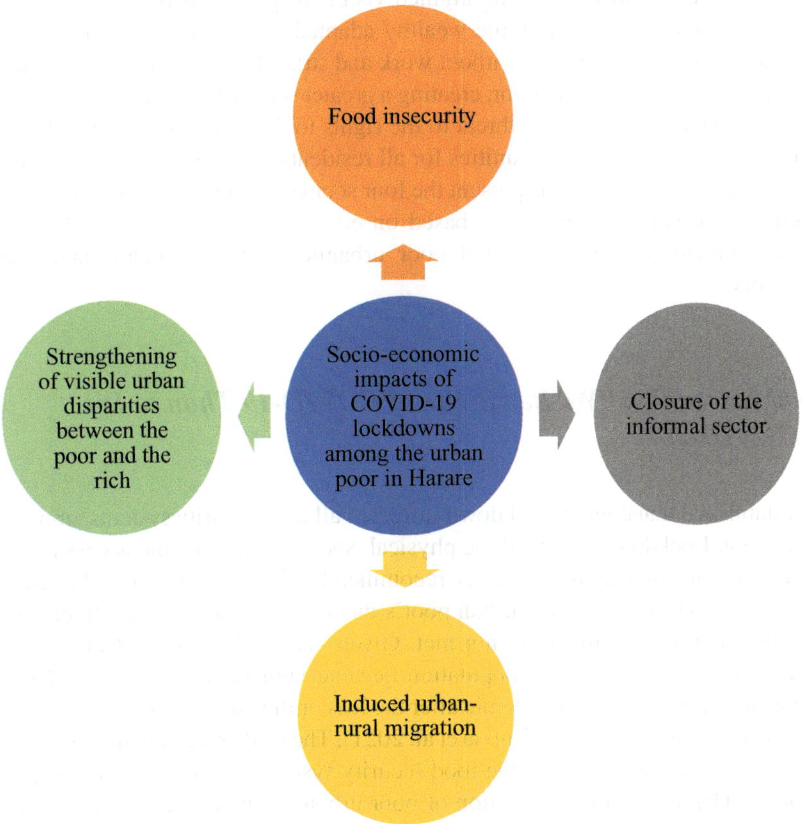

Fig. 5.1 Socioeconomic impacts of COVID-19 lockdowns among the urban poor in Harare. (Source: Authors)

for example, the COVID-19 lockdowns led to widespread job losses, particularly in the informal sector, where many poor urbanites work. The loss of income made it difficult to afford necessities such as food, resulting in increased food insecurity. As a result, food insecurity became a clear manifestation of marginalization and continued suffering of the urban poor during the COVID-19 pandemic. The informal sector, a crucial income source for many poor urbanites, was heavily affected by COVID-19 lockdowns. Millions of marginalized urban population groups suffered a loss of income with the closure of small businesses and informal markets. This resulted in the urban penalty, where poor urbanites are more likely to face economic and social hardship due to their location in urban areas. The COVID-19 mandatory lockdowns induced urban–rural migration as people searched for new sources of income and better living conditions. Many urbanites lost their jobs and struggled to survive in urban areas. They moved to rural areas, where the cost of living was lower and where they could potentially find work in the agricultural sector. This migration marginalized and threatened the rights to the city of the urban poor, who were left behind in cities and faced increased competition for scarce resources.

The COVID-19 lockdowns highlighted visible disparities between the poor and the rich in urban areas. While the wealthy adapted to the lockdowns and worked from home, the poor were left without work and struggled to survive. This widened the gap between the rich and poor, creating a greater sense of marginalization among the urban poor. This is a clear threat to the rights to the city, which demands equal access to resources and opportunities for all residents, regardless of their economic status. The following sections present the four socioeconomic impacts of mandatory lockdowns in Harare, Zimbabwe, based on empirical evidence gathered through remote ethnography from selected poor urbanites, key informants, and online platforms.

Food Insecurity: "We Better Die of COVID-19 Than to Die of HUVID-20"

We established that country lockdowns affected all food security systems among the urban poor. Lockdowns strained the physical, social, and economic access to sufficient, safe, and nutritious food. As recommended by the Food and Agriculture Organization (FAO 2013), the urban poor's dietary needs and food preferences for an active and healthy life were not met. Given that livelihoods in Zimbabwe are largely informal, most of the population became vulnerable in the face of tragic events like the COVID-19 pandemic after the loss or illness of a main income earner or the inability to go to work (Chirisa et al. 2021). The study revealed that COVID-19 lockdowns negatively affected the food security systems of vulnerable urban populations in Harare. A large proportion of poor urbanites in Harare source their food through the informal economy, which was banned during lockdowns. Lockdowns prohibited their main source of income to buy food, while at the same time, there were breaks in food supply chains, food shortages, and basic food price spikes,

Fig. 5.2 Picture aptly showing what average citizens were facing during the lockdown. (Source: WhatsApp Group for Vendors March 2020)

which only appeared in formal retail markets (Muvhuringi et al. 2021). The food security situation among the urban poor worsened due to preexisting shocks before the COVID-19 outbreak. Gumbu (2020) observed that most poor Harare urbanites were already on the extreme end of the hunger spectrum and badly equipped to fend for themselves during COVID-19 lockdowns.

Our remote ethnography discussions with poor urbanites in Harare reflected that food insecurity quickly threatened household well-being. The following animated cartoon (Fig. 5.2) shared by one vendor in a WhatsApp group aptly described what average citizens in Harare were facing during the lockdowns: on the one hand, the real danger of contracting COVID-19 outside, while on the other hand, the reality of hunger for staying indoors. The animated cartoon was accompanied by a message stating that "a new virus has been found in Zimbabwe, discovered on Monday, 30th March 2020; it is called Hunger Virus (HUVID-20), it is causing untold pain to people."[7]

Such dynamics also unfolded among socially segregated populations among India's poor, who found dying of COVID-19 better than dying of hunger (Mishra and Rampal 2020). A clergyman for a local church whom we interviewed telephonically stated that indeed, for poor urbanites, their world was under siege like the biblical times of Samaria,[8] where the famine reached the extent of people eating the

[7] WhatsApp group discussion with female vendors April 2, 2020.

[8] New King James Version Bible, 2 Kings 6:24–33 [[24] Now it came about after this, that Ben-hadad king of Aram gathered all his army and went up and besieged Samaria. [25] There was a great famine in Samaria; and behold, they besieged it, until a donkey's head was sold for eighty *shekels* of silver,

dung of doves and donkeys and even considered eating their children.[9] The majority of the poor urbanites who virtually attended our WhatsApp focus group discussions indicated that they could not afford to sustain their families during lockdowns. Many child- and female-headed households faced difficulties sustaining their families with food during lockdowns. Mrs. Kandemberi, a poor widow who was stopped by lockdown to operate a vegetable stall in Harare CBD, narrated:

> I can't afford to buy food! Groceries are very expensive! We do not have mealie-meal. Cooking oil is very expensive, a bar of soap is now $US4.00. Where do we get all those groceries to sustain us during the lockdown? I survive by selling vegetables in the city, but we have been told to stop selling with mediate effect. I do not have any other source of income. Surviving the 21 days of lockdown will be a struggle.[10]

In a related discussion, a 60-year-old woman, Mrs. Dube, who ran a mini-fast food outlet in Mbare Musika, Harare, expressed how the sudden ban on vending in the city threatened her food security situation with her family. The woman narrated:

> We will die of hunger! Imagine, at my age, I am still working to feed my family. During this first week of lockdown, I failed to provide for the seven kids I am looking after because we cannot get money without doing business. If you cannot even afford a mere COVID-19 mask, what more about putting food on the table?[11]

Food insecurity affected not only the elderly in the city but also the youths and middle-aged who depended on hustling[12] in the city for survival. While most poor urbanites were well-informed about COVID-19, they were radical about adhering to national lockdowns as they had no food. Nowhere was this clearer than expressed in an interview with a 25-year-old man, Gift Kachasu, who survived selling fruits in a cart in Harare's CBD. The young man narrated, "*hapana kusiri kufa* meaning "dying of Corona and dying of hunger it is all death!" That is the truth. I survive by buying and selling; therefore, I should move out of the house to sell.[13]" Another participant commented on Gift's case: "we have multiple responsibilities. Just imagine, on the first day of the lockdown, I wake up with no electricity or water; please tell me how one can practice personal hygiene in such a situation?"

Several participants were skeptical of other social consequences associated with lockdown regulations. For instance, a female money changer in the WhatsApp group lamented that "the lockdown will result in malnutrition among children and plummeting health among the population living with HIV (PLWH) due to lack of

---

and a fourth of a kab of dove's dung for five *shekels* of silver [28] And the king said to her, "What is the matter with you?" And she answered, "This woman said to me, 'Give your son that we may eat him today, and we will eat my son tomorrow.'"]

[9] Key informant interview with Pastor Maonera in Harare, March 28, 2020.

[10] WhatsApp focus group discussion (message posted as voice note) Harare, March 30, 2020.

[11] WhatsApp focus group discussion with informal sector participants, Harare, March 30, 2020.

[12] Hustling typically refers to a way of making a living by engaging in informal or unregulated economic activities, involving low-paying or menial jobs. These may include street vending, informal labor, or other types of work that are often seen as outside of the formal economy.

[13] WhatsApp focus group discussion (message posted as voice note), Harare, April 2020.

nutritious food among the poor urban population."[14] This resonates well with the findings of Dudzai and Wamara (2021). They reflected that food insecurity associated with lockdown restrictions left many urbanities in the informal sector exposed to multiple social challenges such as poor access to health facilities, no finances to access medical resources, increased crime, and gender-based violence. In developing countries like Nigeria and Uganda, food insecurity associated with lockdowns forced the population living with HIV to skip meals or starve as a coping strategy, leaving many depressed (Folayan et al. 2022; Wagner et al. 2022). Youths even faced food insecurity challenges with household responsibilities, indicating they failed to provide food for their families during lockdowns. For example, one of the married youths, Joyline Mazvimbakupa, hinted, "I cannot stay at home watching my two-year-old child dying of hunger because I am scared of coronavirus."[15]

During lockdown, the call for the government to support poor urbanites with food and other essentials was widespread among all WhatsApp focus group discussions. A social media activist highlighted how in other countries like the United Kingdom, the government put in place stimulus packages to preserve citizens' social welfare. Still, the Zimbabwean government did nothing.[16] While the Zimbabwean government claimed it was feeding some of the poor masses, one leader for local residential suburbs revealed that food assistance was patchy and did not replace the daily earnings of poor people who depended on the informal sector for survival.[17] Virtual interviews with other poor urbanites revealed that, though their hopes rested on government to survive the COVID-19-induced lockdowns, government assistance was being distributed through political-related channels, making it inaccessible to everyone regardless of need. Mrs. Romberai narrated that *hapana munhu anofarira kufa mwana wamai*, meaning "no one wants to die of coronavirus my brother but we have no option because powerful political elites members are hijacking government food aid support in our localities."[18]

Closely related to this narration, another youth reflected that "when I went to buy subsidized mealie meal I failed because first preference was being given to ZANU–PF youths and security personnel, i.e., police who were buying more than five bags and later resell to us on higher price."[19] Such irregularities in access to food reflect the ongoing politicization of humanitarian aid in rural areas where perceived supporters of opposition parties are excluded from receiving government and humanitarian assistance even if they meet vulnerability criteria (Munyanyi 2005; Chamunogwa 2021).

The government unveiled a Z$18 billion grant meant for economic recovery and a stimulus package to cater for individuals, families, small business enterprises, and

[14] WhatsApp COVID-19 diary, Harare, July 2021.

[15] WhatsApp focus group discussion, 1 April 2020.

[16] Social media activist on Twitter.

[17] Interview with a leader for a suburb association.

[18] Interview in Mabvuku, during participant observations, Harare, June 2020.

[19] Interview with Mrs. Romberai in Harare, 1 April 2020.

industries. However, during our research, none testified of receiving any allowance from the government. Instead of relying on government support, some urbanites depended on local nongovernmental organizations and charity organizations for support. For instance, an urbanite female cross-border trader in Hatcliffe suburb recalled:

> On the 12th of April 2020 I received a pack of soya chunks, cooking oil and 5kgs of mealie meal from a local Muslim Church. I am so grateful they managed to do what our government is failing to do. They promised that they would return.[20]

Some of the discussions were well-reflected in a short video posted on New Zim TV, titled "We will die of hunger" Despair as Zimbabwe lockdown extended,[21] where several participants from various low-density suburbs testified about the difficulties they were going through to access food following lockdown measures. Therefore, overcoming COVID-19-related deaths was a huge challenge for food security in low-income countries. In this regard, Dzawanda et al. (2021) recommended that to promote sustainability of food production, distribution and accessibility, Zimbabwe should not have blocked roads that linked rural spaces (food producers) with urban areas (food buyers) as part of the lockdown measures. We, therefore, virtually discussed how COVID-19-induced lockdowns affected the rights to the city by the poor and, in essence, the informal sector, which was the primary source of livelihood for the urban poor in Harare.

Closure of the Informal Sector

We observed that COVID-19 extended lockdowns were a big blow to the informal sector in Harare, where most poor urbanites made a living. In Zimbabwe, the informal sector, once derided as an exclusive presence for the uneducated and unskilled, with no prospects of gaining a job in the formal sector, became a lifeline for a growing number of Zimbabweans who were retrenched professionals, retirees, and even highly skilled workers. In essence, the informal sector has become the lifeblood of the Zimbabwean economy. But, COVID-19 became a major economic and labor market shock, presenting significant impacts in terms of unemployment for informal sector workers as all their operations were halted through mandatory lockdowns. Figure 5.3 shows deserted informal sector hawkers' stalls in Harare, Zimbabwe, during lockdown in a bid to slow down the spread of the coronavirus.

Informal sector operators who were affected the most by mandatory lockdowns were the youths who used to operate as transport touts and drivers, fixing/buying and selling mobile phones, money exchangers, small-scale farmers, street vendors, small food retailers, and beauty cosmetologists (see also Dudzai and Wamari 2021;

---

[20] Interview with Clara Zosemesa, COVID-19 dairies in Harare, April 21, 2020.

[21] New Zim Tv "We will die of hunger" Despair as Zimbabwe lockdown extended (https://www.youtube.com/watch?v=rAs-Y3bjr18).

Fig. 5.3 Hawkers' stalls in Harare, Zimbabwe, lie deserted following national lockdown. (Source: WhatsApp Group for Vendors, April 12, 2020)

Dzawanda et al. 2021). Harare urbanites revealed touching stories about how they lost their main source of livelihood following the closure of informal trading during the COVID-19 lockdowns. During a virtual focus group discussion, Mr. Mawoneke a welder stated that

> I rely on welding building materials…but for now, everything is at a standstill, borders are closed, I have no other job…with me here, I have stocks of unsold doors and window frames…who can buy the items under the prevailing situation?[22]

Another female participant commented to Mr. Mawoneke's chat that "building materials are better because they are unperishable…I have lost stock worth USD 50.00, which includes five buckets of tomatoes and ten bundles of vegetables, all rotten, following the closure of our vegetable vending stalls at Mbare Musika."[23] The mandatory directive to cease operations of nonessential activities reduced the mobility of people and goods in the city, significantly affecting the private and informal transport sector, which employed some of the youths in Harare. A virtual

[22] WhatsApp focus group discussion, June 2020.

[23] WhatsApp focus group discussion with vendors, Harare, June 2020.

discussion with commuter omnibus operators revealed that COVID-19 lockdowns doubled poverty levels among youth urbanites in Harare.

One of the youths, Mr. Maodzeke, who operated a commuter omnibus on the Mabvuku suburb to Harare CBD route, reflected that "I used to earn US$200.00 per month, and this was enough to pay for my bills (electricity, rent and water) and to buy food for my family…but with the banning of commuter omnibus operations during lockdowns, tell me how I can get this money? I already owe my landlord US$150 in rentals. I am now resorting to cooking with firewood because I cannot afford electricity bills."[24] During the same discussion, Kelvin who was a tout commented that "*isu mahwindi* 'us touts' we are now stranded because commuters are not operating, we are now spending time-consuming cheap and toxic drugs like *mutoriro* because we cannot afford real beer."[25] Likewise, Dudzai and Wamara (2021) observed that several youths were left jobless following the closure of the informal sector and resorted to criminal activities for survival.

The mandatory closure of international borders negatively affected the informal sector. This affected women, the country's largest proportion of informal cross-border traders. These women own informal sector stalls in Harare CBD and the residential locations where they sell the products they purchase abroad. In her introduction, Mary, a seasoned cross-border trader, stated that "I am a cross-border trader, I sell clothes, footwear and blankets but now I cannot go anywhere because the borders are closed…now you have to present a valid reason, beyond just going to buy stuff for selling, to the Department of Immigration if you want to cross the border."[26] Commenting on this Tsitsi, another female cross-border trader openly expressed that "the road border is for us who are poor, those with money are flying to China, South Africa, Dubai, and Turkey to buy stuff for reselling…regardless of COVID-19 lockdowns…. it's us who are poor who will die of poverty during this COVID-19."[27] Another female participant recorded that "my mother passed away three weeks after the lockdown was announced…I was the only one who could afford to buy her medication, but I was delayed at Beitbridge Border Post by immigration officers stating that I must be quarantined for two weeks."

Regardless of the government and local authorities being aware that poor urbanites depend on the informal sector for daily sustenance, state security continued to disperse informal sector operators who opened their businesses during the lockdowns. On several occasions from March to April 2020, police and soldiers were deployed to forcefully displace informal traders violating the COVID-19 lockdown regulations. A vendor operating in Mabvuku audio recorded in a WhatsApp focus group discussion that "today the police and soldiers were deployed to disperse us 'vendors' operating in the business centers in this location…some vendors were

[24] WhatsApp discussion with youths, Harare, March 2021.

[25] WhatsApp discussion with youths, March 2021.

[26] Voice note sent on WhatsApp discussion with informal traders, Harare, April 2020.

[27] WhatsApp focus group discussion with informal traders, Harare, April 2020.

beaten, and goods were destroyed during the operation."[28] Similarly, another informal trader, Jairos, who operated a small gas shop in Hatcliffe, stated that "The riot police severely tortured me after they found my tuckshop opened. The experience was very traumatizing."[29] One civic activist on Twitter's social media warned that "in other countries, the police have reportedly already killed more violators of the lockdown than people have died from the virus. We don't want that same experience in Zimbabwe."[30] Apart from being brutal, state security agents were also accused of demanding money or favors from some informal sector operators to allow them to continue operating during the lockdown. Cosmas, a commuter omnibus driver on the Kuwadzana/Harare CBD route, revealed that "the only way to survive in business is to pay the traffic police…I pay at least USD30.00 to them so that I can operate the whole day without being disturbed.[31]" Another vendor revealed that "business is difficult to conduct…whenever I am caught by police or council police, I will pay USD 20.00 so that I will not be fined USD50.00."[32]

During our transect drives around Harare CBD and the surrounding suburbs, we observed that the local authority took advantage of lockdown powers to go on a demolition spree, displacing small vendors' stalls en masse. The justification behind the demolitions was that the structures had become an eyesore and hubs of the COVID-19 virus. Some of the vendors were left with no means to resume operations even after the lifting of lockdown. In Mabvuku, for instance, we witnessed some vendors selling their items on the roadside after the council demolished their tuckshops. In the Hatcliffe suburb, Mr. Chikwara, a grinding mill operator, revealed that "my grinding mill was left on open when the cabin was destroyed…I have no option but to sleep here guarding it."[33] Although the council assured the building of new market complexes in designated vending zones, vendors feared that the new stalls were not allotted with urgency. Figure 5.4 shows a picture shared in a WhatsApp group for vendors when the City of Harare local authority demolished makeshift structures set up by vendors during the mandatory COVID-19 lockdown.

In short, the rich remote ethnographic experiences among Harare urbanites revealed that COVID-19 mandatory lockdowns became an assault on the urban poor's livelihoods and imagined futures, which were wholly dependent on the informal sector. Women, youth, and children, who constituted the greater percentage of the informal economy, experienced the greatest shocks from mandatory lockdown measures. This situation penalized vulnerable groups as they disproportionately experienced the negative consequences of closing the informal sector through mandatory lockdown policies. Furthermore, the closure of the informal sector further

[28] COVID-19 Diary, May 2020.

[29] COVID-19 Diary, Harare, May 2020.

[30] WhatsApp text message by a civic activist during an active COVID-19 discussion in a WhatsApp group April 15, 2020.

[31] WhatsApp focus group discussion with youths, Harare, June 2020.

[32] WhatsApp focus group discussion with vendors, Harare, March 2021.

[33] Interview during observations exercise, Harare, July 2020.

Fig. 5.4 Harare City Council demolishing the urban poor's vending stalls during COVID-19 lockdown. (Source: WhatsApp Group for Vendors, April 17, 2020)

marginalized the poor, limiting their ability to access informal networks and resources that could help them cope with economic and social challenges during the pandemic. This threatened the rights to the city, as the urban poor faced barriers to accessing basic services, participating in decision-making processes, and enjoying equal opportunities of accessing city spaces for economic and social advancement. To address these challenges, it is important that governments and local authorities support the informal sector and ensure that the needs of the urban poor are prioritized in policy and planning initiatives in times or crises like the COVID-19 pandemic.

Induced Urban–Rural Migration

The COVID-19 mandatory lockdowns triggered urban-to-rural migration, relegating some poor urban dwellers back to rural areas. The banning of the informal sector, where the majority of poor urbanites made a living, meant some could not survive in the city during the lockdown. Ahead of the first 21-day Zimbabwean COVID-19 mandatory lockdown, we observed that thousands of people thronged Mbare Musika bus terminals in Harare and many other pickup points seeking transport to various destinations in rural areas. Figure 5.5 shows Zimbabweans migrating to rural areas ahead of the first 21-day lockdown.

Rapid interviews were carried out at bus terminals soon after the announcement of the first mandatory lockdown revealed that the majority chose to go to their rural areas during lockdown to stay with relatives as they could not survive in the city

Fig. 5.5 Zimbabweans fleeing to rural areas ahead of first COVID-19 21-day lockdown. (Source: WhatsApp Group for Youths, March 18, 2020)

without an income. One traveler, Gertrude Chamboko, said she would rather observe the lockdown days in Gutu, her rural home, than in the city without a proper income and squeezing herself in one room with her family. She narrated:

> I am heading home to Gutu. This is where I will be for the next 21 days. I should spend the stipulated time there with my family because we survive on buying and selling in the informal sector, which the government has suddenly banned.[34]

Another traveler, Simon Ponde, at Mbudzi Roundabout, an illegal pickup point where other hundreds of travelers were trying to find any mode of transport to their rural destinations, narrated:

> I am going home to Masvingo with my family. However, I have been here for the past three hours, and not even a bus has passed. That is why there are a lot of people here. What is now worrying is that transporters have hiked fares. We used to pay US$8 to go to Masvingo from Harare, but now they ask for US$20 to US$30. We have no choice but to spend the three-week shutdown in the village, where we have better chances of getting food and water supplies with minimal possibility of high prices.[35]

In another similar case, a traveler, Joseph Mapuranga, whose company had closed because of the lockdown, had this to say:

> It is not easy to migrate to Chiredzi with your whole family. My company has closed, and I am unsure if we will receive this month's salary. I decided to move to my rural home in Chiredzi so my children would not suffer from hunger. I cannot survive this lockdown

[34] Interview with Getrude Chamboko in Harare, 29 March 2020.

[35] Interview with Simon Ponde in Harare, 29 March 2020.

> period without any financial help. I live in a one-roomed house with my three children, and keeping social distance might not be easy; hence I am opting for my village.[36]

Commenting on the massive mobility of people from urban to rural areas during the COVID-19 lockdown, Robson Maura, a driver at Mandaza buses, confirmed that transport operators recorded brisk business as people jostled to get transport to their respective rural homes. Robson's company had to increase the number of buses on most rural routes to cater for the increased number of passengers. He narrated:

> Today, we had seven buses going to Chiredzi, all full. People are fleeing the city, saying there is coronavirus, and they cannot afford to survive 21 days without working. They should be in rural areas. On our side, we are happy because we have managed to capitalize on our profits.[37]

Fears, however, abounded that the urban–rural migration could spread COVID-19 to the country's already marginalized rural communities. The logic of country lockdown was that symptoms developed after 14 days thus should have manifested within the 21-day lockdown period. Tracking contacts was easy, but the whole village might have been infected by people returning to rural areas without knowing their status. One health expert, Dr. Daimon, narrated saying:

> It is quite unfortunate that people are massively migrating to rural areas. If they spread the disease there, it will be a big disaster countrywide, as the government cannot contain the virus. Unfortunately, people decided to migrate to rural areas, which would be a disaster. We have been urging the government to equip all hospitals with the protective measure for Coronavirus, but it did not listen. People in rural areas do not have information on how to protect themselves from infected persons. You cannot greet your old parents with an elbow; they do not understand that.[38]

Clearly, COVID-19 lockdowns generated new mobility dynamics (Bhanye et al. 2021), with the urban poor relegated out of the city. This contradicts popular literature wherein; mobility is dominantly characterized by the movement of people from rural to urban areas, not the other way around (Mlambo 2018; De Brauw et al. 2014). High rates of rural-to-urban migration account for more than half of the growth of African cities (Njwambe et al. 2019). While migrating to villages was visible, we encountered several youths migrating to emerging peri-urban centers with unrestricted economic activities. For instance, during a WhatsApp discussion, James (a cross-border trader) posted a photo of his family in a truck with their belongings, relocating to Domboshava. He also added a voice note stating that "living costs in Harare are now unaffordable…due to COVID-19 lockdowns and border closure…I am therefore relocating to Domboshava where I can easily access mining sites in Mazowe and Bindura."[39] Mining communities were targeted by migrating urbanites during the COVID-19 lockdowns. During the same discussion, another female participant Bertha asked James (an illegal gold panner in Mazowe) if he

---

[36] Interview with Joseph Mapuranga in Harare, 29 March 2020.

[37] Interview with Robson Maura in Harare, March 29, 2020.

[38] Interview with Health Expert Dr. Daimon in Harare, March 29, 2020.

[39] Focus group discussion on WhatsApp, April 13, 2020.

would assist her in finding a space to open a stall in Mazowe to sell vegetables, clothes, and beer since her previous stall in Harare CBD was inaccessible due to the COVID-19 lockdowns.[40] Countrywide, for instance, in the Eastern Highlands of Zimbabwe, national lockdowns and border closures left many youths without viable economic options, except for illegal gold mining in protected areas.

Strengthening of Visible Urban Disparities Between Poor and the Rich

Apart from the negative impacts of COVID-19 mandatory lockdowns on the urban poor, we observed the strengthening of visible urban disparities between the poor and the rich in Harare. Social media posts on Twitter and Facebook revealed that the rich and elite endorsed country lockdowns without hesitation. At the same time, the urban poor wondered how they would survive the lockdowns. Through discussions with participants, we noticed disparities in purchasing power between the middle class, urban elites, and urban poor following lockdown pronouncements. The middle class especially those employed in the private sector working online, essential service providers, and the elite besieged supermarkets and shops in the cities and towns for groceries and other essentials needed for the lockdown period. Despite noticeable price increases, there were long queues in shops in Harare as the privileged bought basic products. Figure 5.6, shared by one of the participants, shows the rich and elite queuing to buy groceries as they braced for the first 21-day lockdown in Harare.

A vendor selling near an OKmart supermarket witnessed affluent customers congesting the parking area and forming long queues to purchase key household items, which included cartons of rice, soap, pasta, spaghetti, flour, mealie meal, beans, matemba, powder milk, breakfast cereals, dried meat, and fish. Other kitchen supplies included salt, sugar, bottled dry spices, tomato sauce, and cooking oil. During the same discussion, Mrs. Mamoyo a hairdresser, shared that she was given some masks, paracetamol, and gloves by one of her clients from Borrowdale, who had stocked medication items, sanitary ware, and food supplies. Similarly, Tinotenda, who used to be a commuter driver, reflected that while his family was resorting to firewood for energy and plastic materials following shortage and increase in the price of gas and electricity, his employer (the commuter owner) had purchased in bulk prepaid electricity units, bought extra gas, charcoal, torches (chargeable and solar), torch batteries, candles, matches, lighters, generators, and extra fuel for generators and cars.

Conversely, we observed most of the urban poor struggling to stock up on enough food and other utilities ahead of the lockdown. Throughout the lockdown, most of

---

[40] WhatsApp focus group discussion, May 2020.

Fig. 5.6 The elite and middle-class queue to buy groceries as they brace for 21-day lockdown. (Source: WhatsApp Group for Vendors, March 26, 2020)

the urban poor relied on subsided food items; however, everyone stampeded for the items.

The gaps in the purchasing power of urban populations in Harare reflected the strengthening of visible urban disparities between the poor and the rich during COVID-19 mandatory lockdowns. With the COVID-19 pandemic, poverty levels increased and became more pronounced in urban areas and among female-headed households and the youth. But, even before the COVID-19 pandemic, poverty and inequality increased in countries in crisis like Zimbabwe (OSISA 2021). The distribution of resources and income in Zimbabwe was highly skewed, and the levels of income inequality remained among the highest in the world.

The wealth gaps had grown because of the absence of systematic transformation of economic structures and the perpetuation of a typical African enclave economy. Here the economy is characterized by a small and well-resourced formal sector, which operates in isolation from a large, growing, and poverty-stricken informal economy and communal subsistence economy (OSISA 2021). As a result, Zimbabwe can be described as a country in deep crisis. The widening of the already existing disparities by the COVID-19 pandemic increased the sense of injustice and deprivation of the urban poor in Zimbabwe. According to recent estimates, the wealthiest 10% of the population consumes nearly 40% of national resources (Fight Inequality 2022). This skewed distribution of resources and economic opportunities is detrimental to national cohesion and the achievement of national objectives. There is, thus, no doubt that Zimbabwe needs a fresh and different approach to address the current socioeconomic inequalities perpetuated by the COVID-19 pandemic. Without proper government intervention, the visible urban disparities between the poor and the rich could continue growing beyond the COVID-19 pandemic. Recent studies suggest the importance of having a strong middle class that can exert pressure on the government for democratization and mobilization for reform in Zimbabwe (Chikwaza and Chikumbu 2021).

Conclusion: Deepened Social and Spatial Inequality and Threats to the Right to the City of the Urban Poor

In this chapter, we observed and discussed four major socioeconomic impacts of COVID-19 mandatory lockdowns on the urban poor in Harare, Zimbabwe: food security systems of poor urbanites who source food through the informal economy, closure of the informal sector, induced urban–rural migration, and strengthening of visible urban disparities between poor and the rich. From the findings, we conclude that the COVID-19 pandemic mandatory lockdowns deepened social and spatial inequality among the urban poor, threatening their right to the city. The socioeconomic impacts upsurged urban poverty, increased unemployment and the risks of hunger and food insecurity, reinforced existing inequalities, and broke social harmony in the cities, even past the COVID-19 pandemic period. While some studies demonstrated positive impacts of COVID-19 lockdowns like slowing the spread of the virus (e.g., Lau et al. 2020 in China) and economic benefits like people working remotely from home (Atalan 2020), in the global north; our findings reflect similar dimensions with other African studies.

These African studies reflect the negative socioeconomic impacts of the COVID-19 pandemic lockdown on the urban poor, ranging from disruption of the informal sector, loss of jobs, forceful relocations, increased food prices, and stress and boredom (Carmody et al. 2020; Chirisa et al. 2020; Asante and Mills 2020; Matamanda et al. 2022; Mulugeta et al. 2021). Urban scholars agree that COVID-19 mandatory lockdowns caused asymmetrical spatial impacts, magnifying preexisting divisions within cities (Matamanda et al. 2022; Mulugeta et al. 2021; Turok and Visagie 2021). Townships and informal settlements proved more vulnerable than other urban spaces (Turok and Visagie 2021). From a right-to-the-city perspective, these socioeconomic impacts must be considered to make just cities for all. In 2016, Habitat III (UN Conference on Housing and Sustainable Urban Development, which took place in Quito, Ecuador) resulted in a New Urban Agenda, which endorsed the ideas of the rights to the city, calling on urban development to "leave no one behind" (UN 2021). The findings presented in this chapter contribute to the debate on urban citizenship and the right to the city movements by demonstrating how the COVID-19 pandemic revealed the failures of current urban governance models.

Our study was unique because of the adoption of new and agile directions in research, ethnography, to engage marginalized urbanites to tell their stories in detail on the socioeconomic impacts of COVID-19 mandatory lockdowns on everyday lives in Harare, Zimbabwe. Our remote ethnography generated rich and powerful insights from participants through texts and video diaries and participant photo uploads. We found remote ethnography more cost-effective, less intimidating, easier for respondents, and faster because of in-the-moment feedback. Unlike traditional full immersion ethnography, remote ethnography enabled us to deal with the complexities of the "global," the "local," and the "trans-local" nature of modern urban research participants. With difficulties in mobility due to COVID-19 restrictions for researchers and wider populations, we suggest that it is time for researchers in the Global South to utilize remote ethnography methodologies. Remote ethnography is nimble and cuts across geographical boundaries and societal structures of power, unlike traditional research approaches. Some criticisms have been levelled against remote ethnography, for example, the need for the remote ethnographer to possess certain technology-based skills, the temporal nature of online data, and issues of privacy (Kaur-Gill and Dutta 2017); however, with skill, flexibility and reflexivity in their practice, remote ethnographers can successfully utilize the innovative methodology.

References

Adams-Prassl A, Boneva T, Golin M, Rauh C (2020) Inequality in the impact of the coronavirus shock: evidence from real-time surveys. J Public Econ 189:104245. https://doi.org/10.1016/j.jpubeco.2020.104245

Asante LA, Mills RO (2020) Exploring the socio-economic impact of COVID-19 pandemic in marketplaces in urban Ghana. Afr Spectr 55(2):170–181

Atalan A (2020) Is the lockdown important to prevent the COVID-19 pandemic? Effects on psychology, environment and economy-perspective. Ann Med Surg 56:38–42

Avetisyan S (2020) Coronavirus and urbanization: does pandemics are anti-urban? Available at SSRN 3584395

Balde R, Boly M, Avenyo E (2020) Labour market effects of COVID-19 in sub-Saharan Africa: an informality lens from Burkina Faso, Mali and Senegal (UNU-MERIT working paper 022). Maastricht Economic and Social Research Institute on Innovation and Technology (UNU-MERIT)

Bhanye J (2022) "Lydiate is now our home of a sort": perceptions of place amongst ageing first-generation Malawian migrants in Zimbabwe. Anthropol South Afr 45(3):180–194

Bhanye J (2023) "Nimble Sociality and Belonging": an Ethnography of Migrants' Responses to Bans on Associational Life During the COVID-19 Pandemic. In Urban Forum (pp. 1–27). Dordrecht: Springer Netherlands. https://doi.org/10.1007/s12132-023-09503-0

Bhanye J (2023a) 'Emerging forms of spatialised and socialised authority' among 'tenure-insecure peri-urbanites' in African peri-urban spaces: a review study. Pan-Afr Conversations 1(2):26–63

Bhanye J (2023b) "Emerging forms of authority in land access?": the occult and witchcraft among Malawian migrants in peri-urban Zimbabwe. In: Goździak EM, Main I (eds) Debating religion and forced migration entanglements. Springer, Cham, pp 89–107

Bhanye J, Bhanye A (2023) A blessing or a curse?: the role of social media during the COVID-19 pandemic in Africa. In: The Palgrave handbook of global social problems. Palgrave Macmillan, Cham. https://doi.org/10.1007/978-3-030-68127-2_367-1

Bhanye J, Dzingirai V, Chirisa I (2021) Transnational migrants on the margin: agency, aspirations, and perceptions for the future among Malawian migrants in Zimbabwe's Norton Peri-

Urban. In: Brears RC (ed) The Palgrave encyclopedia of urban and regional futures. Springer International Publishing, Cham, pp 1–10

Bhanye J, Shayamunda R, Tavirai RC (2023) Social media in the African context. A review study on benefits and pitfalls. In: The Palgrave handbook of global social problems. Palgrave Macmillan, Cham. https://doi.org/10.1007/978-3-030-68127-2_366-1

Bhardwaj G, Esch T, Lall SV, Marconcini M, Soppelsa ME, Wahba S (2020) Cities, crowding, and the coronavirus. World Bank policy research working paper (9278)

Carmody P, Mccann G, Colleran C, O'halloran C (2020) COVID-19 in the Global South: impacts and responses. Bristol University Press, Bristol

Chamunogwa A (2021) The impact of COVID 19 on socio-economic rights in Zimbabwe. The Zimbabwe Peace Project. 5(10):127–8

Chikwaza D, Chikumbu A (2021) Bridging the gap between rich and poor: a strong middle class is key to achieving democratic transformation in Zimbabwe. Retrieved from https://www.dailymaverick.co.za/article/2021-05-06-bridging-the-gap-between-rich-and-poor-a-strong-middle-class-is-key-to-achieving-democratic-transformation-in-zimbabwe/. Accessed 1 Sept 2022

Chirisa I, Mutambisi T, Chivenge M, Mabaso E, Matamanda AR, Ncube R (2020) The urban penalty of COVID-19 lockdowns across the globe: manifestations and lessons for Anglophone sub-Saharan Africa. GeoJournal 87:1–14

Chirisa I, Mavhima B, Nyevera T, Chigudu A, Makochekanwa A, Matai J, Mundau L (2021) The impact and implications of COVID-19: reflections on the Zimbabwean society. Soc Sci Humanit Open 4(1):100183

Clifford J, Marcus GE (eds) (1986) Writing culture: the poetics and politics of ethnography: a School of American Research advanced seminar. University of California Press, Berkeley

De Brauw A, Mueller V, Lee HL (2014) The role of rural–urban migration in the structural transformation of sub-Saharan Africa. World Dev 63:33–42

Debuquet DL, Martin W, Vos R (2020) Impacts of COVID-19 on global poverty, food security and diets, IFPRI discussion paper 01993. International Food Policy Research Institute, Washington, DC

Dudzai C, Wamara CK (2021) COVID-19 pandemic and the informal sector in Zimbabwe. African Journal of Social Work. 11(4):201–8

Dzawanda B, Nicolau MD, Matsa M (2021) Impact of virtual cash economy on livelihood outcomes of informal cross border traders in Gweru, Zimbabwe. In Urban Forum. 32(4): 521–539. Dordrecht: Springer Netherlands

Espi G, Ranchhod V, Leibbrandt M (2020) The relationship between employment history and COVID19 employment outcomes in South Africa. NIDS-CRAM wave 2 reports, working paper 6

Fight Inequality (2022) 2022 National budget is it a #budget4equality or an instrument for capitalist accumulation? https://www.fightinequality.org/news/2022-national-budget. Accessed 13 Sept 2022

Folayan MO, Ibigbami O, Brown B et al (2022) Differences in COVID-19 preventive behavior and food insecurity by HIV status in Nigeria. AIDS Behav 26:739–751. https://doi.org/10.1007/s10461-021-03433-3

Food and Agriculture Organization of the United Nations (2013) The state of food insecurity in the world 2013. The multiple dimensions of food security. http://www.fao.org/3/a-i3434e.pdf. Accessed 28 June 2022

Gisselquist R, Kundu A (2020) Horizontal Inequality, COVID-19, and lockdown readiness: evidence from India, WIDER working paper 2020/156. UNU-WIDER, Helsinki. https://doi.org/10.35188/UNUWIDER/2020/913-6

Gumbu YL (2020) COVID-19's impact on the livelihoods of informal traders in Zimbabwe. Available at: https://moderndiplomacy.eu/2020/06/16/COVID-19s-impact-on-the-livelihoods-of-informal-traders-in-zimbabwe/. Accessed 16 June 2022

Herald (2020) Breaking: 21 days lockdown to start Monday, 2020, March. Retrieved from https://www.herald.co.zw/breaking-21-days-lockdown-to-start-monday/. Accessed 30 Jan 2021

Hjorth L, Horst HA, Galloway A, Bell G (eds) (2017) The Routledge companion to digital/remote ethnography. Routledge, New York, pp 21–28

ILO (2020) COVID-19 crisis and the informal economy: immediate responses and policy challenges. https://www.ilo.org/wcmsp5/groups/public/@ed_protect/@protrav/@travail/documents/briefingnote/wcms_743623.pdf. Accessed on 30 Jan 2021

Kaur-Gill S, Dutta MJ (2017) DigitalRemote ethnography. In: Matthes J, Davis CS, Potter RF (eds) The international encyclopedia of communication research methods. Wiley, Hoboken, pp 1–10

Kimani-Murage EW, Schofield L, Wekesah F, Mohamed S, Mberu B, Ettarh R, Egondi T, Kyobutungi C, Ezeh A (2014) Vulnerability to food insecurity in urban slums: experiences from Nairobi, Kenya. Journal of urban health. 91:1098–1113

Kumoro NB (2022) Ethnographic methods in the globalised and post-pandemic society: a virtual-collaborative fieldwork. In: ICOLLEC 2021: proceedings of the 1st international conference on language, literature, education and culture, ICOLLEC 2021: 9–10 October 2021, Malang, Indonesia. European Alliance for Innovation, p 62

Laborde D, Martin W, Vos R (2021) Impacts of COVID-19 on global poverty, food security, and diets: insights from global model scenario analysis. Agric Econ 52(3):375–390

Lau H, Khosrawipour V, Kocbach P, Mikolajczyk A, Schubert J, Bania J, Khosrawipour T (2020) The positive impact of lockdown in Wuhan on containing the COVID-19 outbreak in China. J Travel Med 27(3):taaa037

Lawrence L (2022) Conducting cross-cultural qualitative interviews with mainland Chinese participants during COVID: lessons from the field. Qual Res 22(1):154–165

Lone SA, Ahmad A (2020) COVID-19 pandemic – an African perspective. Emerging Microbes Infect 9(1):1300–1308

Mackworth-Young CR, Chingono R, Mavodza C, Mchugh G, Tembo M, Chikwari CD, Ferrand RA (2021) Community perspectives on the COVID-19 response, Zimbabwe. Bull World Health Organ 99(2):85

Makombe EK (2021) “Between a rock and a hard place”: the coronavirus, livelihoods, and socio-economic upheaval in Harare’s high-density areas of Zimbabwe. J Dev Soc 37(3):275–301

Marcus GE (1995) Ethnography in/of the world system: the emergence of multi-sited ethnography. Ann Rev Anthropol 24:95–117

Matamanda AR, Dunn M, Nel V (2022) Broken bridges over troubled waters: COVID-19 and the urban poor residing in Dinaweng informal settlement, Bloemfontein, South Africa. S Afr Geogr J 104:1–19

Matsungo TM, Chopera P (2020) Effect of the COVID-19-induced lockdown on nutrition, health and lifestyle patterns among adults in Zimbabwe. BMJ Nutr Prev Health 3(2):205

Mishra K, Rampal J (2020) The COVID-19 pandemic and food insecurity: a viewpoint on India. World Dev 135:105068

Mlambo V (2018) An overview of rural-urban migration in South Africa: its causes and implications. Arch Bus Res 6(4):32–41

Moran L, Caetano A (2021) Biographical research through the looking glass of social distancing: reflections on biographical interviewing and online technologies in pandemic times. Ir J Sociol 50(2):147–164. https://doi.org/10.1177/07916035211022182

Mulugeta T, Tadesse E, Shegute T, Desta TT (2021) COVID-19: socio-economic impacts and challenges in the working group. Heliyon 7(6):e07307. https://doi.org/10.1016/j.heliyon.2021.e07307

Munyanyi RM (2005) The political economy of food aid: a case of Zimbabwe. Doctoral dissertation. University of the Western Cape

Muvhuringi P, Nyamuziwa B, Kudzai T, Chigede N (2021) The impact of COVID-19 on agricultural extension and food supply in Zimbabwe. Cogent Food Agric 7(1):1918428. https://doi.org/10.1080/23311932.2021.1918428

Mwambari D, Purdeková A, Bisoka AN (2021) Covid-19 and research in conflict-affected contexts: distanced methods and the digitalisation of suffering. Qual Res:1–20. https://doi.org/10.1177/1468794121999014

Njwambe A, Cocks M, Vetter S (2019) Ekhayeni: rural–urban migration, belonging and landscapes of home in South Africa. J South Afr Stud 45(2):413–431

Nyabeze K, Chikoko W (2021) Socio-economic impact of COVID-19 lockdown measures on the informal sector livelihoods in Zimbabwe. Afr J Soc Work 11(4):231–239

Nyanga T, Zirima H (2020) Reactions of small to medium enterprises in Masvingo, Zimbabwe to COVID 19: implications on productivity. Bus Excellence Manage 10(1):22–32

OECD (2020) Cities policy responses. OECD policy responses to coronavirus (COVID-19). OECD, Paris. https://www.oecd.org/coronavirus/policy-responses/cities-policy-responses-fd1053ff/. Accessed 22 Jan 2022

OECD (2021) Addressing COVID-19 and moving to net zero greenhouse gas emissions. https://www.oecd.org/publications/oecd-regional-outlook-2021-17017efe-en.htm

OSISA (2021) Inequality in southern Africa: options for redress. Open Society Initiative for Southern Africa. Policy Brief. https://www.ilo.org/wcmsp5/groups/public/%2D%2D-ed_dialogue/%2D%2D-actrav/documents/meetingdocument/wcms_230181.pdf. Accessed 10 Sept 2022

Paupini C, Teigen HF, Habib L (2022) A change of space: implications of digital fieldwork in connected homes during the COVID-19 pandemic. Digital Creativity:1–15. https://doi.org/10.1080/14626268.2022.2041141

Schotte S, Danquah M, Osei RD, Sen K (2021) The labour market impact of COVID-19 lockdowns: evidence from Ghana. WIDER working paper 2021/27. UNU-WIDER. https://doi.org/10.35188/UNU-WIDER/2021/965-5

Tampe T (2021) Potential impacts of COVID-19 in urban slums: addressing challenges to protect the world's most vulnerable. Cities Health 5(sup1):S76–S79. https://doi.org/10.1080/23748834.2020.1868851

Teachout M, Zipfel C (2020) The economic impact of COVID-19 lockdowns in sub-Saharan Africa. International Growth Centre. https://www.theigc.org/wp-content/uploads/2020/07/Teachout-Zipfel-2020-Working-Paper.pdf

Thurlow J (2020) COVID-19 lockdowns have imposed substantial economic costs on countries in Africa. In: Fan S, Kanbur K, Rahman S, Thurlow J (eds) COVID-19 in developing economies. International Food Policy Research Institute, pp 23–25. https://doi.org/10.2499/p15738coll2.133116

Turok I, Visagie J (2021) COVID-19 amplifies urban inequalities. S Afr J Sci 117(3–4):1–4. https://doi.org/10.17159/sajs.2021/8902

UN (2021) The new urban agenda. https://habitat3.org/the-new-urban-agenda/. Accessed 20 Jan 2022

UN-Habitat (2020a) UN-Habitat COVID-19: key messages. https://unhabitat.org/sites/default/files/2020/03/COVID19_key_messages_eng_1.pdf. Accessed 20 Jan 2022

UN-Habitat (2020b) Opinion: COVID-19 demonstrates urgent need for cities to prepare for pandemics. United Nations Human Settlements Programme. https://unhabitat.org/opinion-covid-19-demonstrates-urgent-need-for-cities-to-prepare-for-pandemics. Accessed 20 Jan 2022

United Nations (UN) (2020) Policy brief: the world of work and COVID-19. https://www.un.org/sites/un2.un.org/files/the_world_of_work_and_covid-19.pdf. Accessed 10 Jan 2022

Varis P (2016) Digital ethnography. In: Georgakopoulou A, Spilioti T (eds) The Routledge handbook of language and digital communication. Routledge, London, pp 55–68

Wagner GJ, Wagner Z, Giza M et al (2022) Increased depression during COVID-19 lockdown associated with food insecurity and antiretroviral non-adherence among people living with HIV in Uganda. AIDS Behav 26:2182–2190

Watson A, Lupton D (2022) Remote fieldwork in homes during the COVID-19 pandemic: video-call ethnography and map drawing methods. Int J Qual Methods 21:16094069221078376

World Bank (2021) Zimbabwe economic update: COVID-19 further complicates Zimbabwe's economic and social conditions. Retrieved from https://www.worldbank.org/en/country/zimbabwe/publication/zimbabwe-economic-update-covid-19-further-complicates-zimbabwe-s-economic-and-social-conditions. Accessed 30 Jan 2022

Chapter 6
From Crisis to Action: Emerging Perspectives and the Morphing of a Sustainable Urban Future Post-COVID-19 Pandemic

Introduction

The COVID-19 pandemic brought about unprecedented disruptions to our daily lives, causing global health, economics, and social systems crises. However, amidst these challenges, it prompted a rethinking of how we live, work, and interact with our urban environments. Before concluding the book, we will give a brief recap of the book. Chapter 2, *Urban Penalty and the Right to the City of the Urban Poor During the COVID-19 Pandemic*, presented the book's theoretical framework focusing on the urban health penalty theory, the rights to the city theory, distributive justice theory, and the complex theory. Chapter 3, *Contextualizing Harare Urban Socioeconomic Profile and History of Pandemics in the City*, presented the socioeconomic profile of Harare, the capital city of Zimbabwe, highlighting the vulnerabilities of the urban poor over time. The chapter discussed pandemics and epidemics that ravaged the city of Harare and how they were intertwined with socioeconomic, geopolitical, and governance issues, which unfolded in the city and country at large. Chapter 4, *The COVID-19 Pandemic and Urban Policy Interventions in Zimbabwe*, focused on the Government of Zimbabwe's policy interventions to address the COVID-19 pandemic. The chapter argued that most COVID-19 pandemic policies, like mandatory lockdowns, were top-down and interventionistic, as they were directives from the WHO, which directed global interventions. Subsequently, these policies had little regard for the welfare of the urban poor resulting in worsened suffering of the urban poor who struggled to make ends meet. Chapter 5 focused on *The COVID-19 Lockdowns and Poor Urbanites in Harare, Zimbabwe: Exploring Socioeconomic Impacts with Remote Ethnography.* While the COVID-19 pandemic presented new challenges to traditional fieldwork research, we innovatively embraced new and agile directions in research methodologies by adopting remote ethnography for this chapter. The chapter illuminated four significant socioeconomic impacts associated with COVID-19 lockdowns among the urban poor: the

J. I. Bhanye et al., *COVID-19 Lockdowns and the Urban Poor in Harare, Zimbabwe*, https://doi.org/10.1007/978-3-031-41669-9_6

food security systems of poor urbanites who consume through the informal economy, closure of the informal sector, induced urban–rural migration, and strengthening of visible urban disparities between the poor and the rich. This final chapter, *From Crisis to Action: Emerging Perspectives and the Morphing of a Sustainable Urban Future Post-COVID-19 Pandemic*, concludes the book, exploring the changing landscape of urban sustainability in the wake of the pandemic. The chapter presents the key issues and emerging perspectives from the book and highlights innovative solutions that can shape a more sustainable and resilient future.

Key Issues and Emerging Perspectives from Harare

Several issues emerged from the book. First, the book "COVID-19 Lockdowns and the Urban Poor in Harare, Zimbabwe: Emerging Perspectives and the Morphing of a Sustainable Urban Future" confirms the history of pandemics in the city of Harare and their impact on vulnerable communities. Like many other cities, Harare has a long history of pandemics, from cholera and typhoid outbreaks to the devastating HIV/AIDS pandemic that disproportionately affected the urban poor. The critical issue emerging from these pandemics is the vulnerability of cities to pandemics. The contagion of cities cannot be ignored as the high-density population and increased social interactions and intercourse provide ripe ground for spread of diseases, especially among the poor. The book explored how past pandemics contributed to the systemic vulnerabilities of the urban poor in Harare and how the COVID-19 pandemic exacerbated these vulnerabilities.

Second, the urban health penalty is apparent and shows how the city fathers and governance system marginalize the urban poor in Harare. The book demonstrates that the urban poor in Harare are more vulnerable to health risks, including COVID-19, due to their living conditions, lack of basic amenities such as clean water and sanitation, and limited access to healthcare. These conditions are nothing new to poor neighborhoods, as the urban health penalty trends of the poor date back to Industrial Revolution cities where the poor lived in squalor and were vulnerable to disease outbreaks. The government's failure to provide basic services such as water, sanitation, and healthcare perpetuated poverty and inequality in the city, simultaneously leading to disparities in health outcomes between high-income urbanites and the urban poor. The book highlights how the COVID-19 pandemic and mandatory lockdowns exacerbated the urban health penalty, increasing infection and deaths among the urban poor.

Third, the COVID-19 pandemic magnified the injustices against the urban poor and the failure to recognize their right to the city. This sadly aligns with the "sanitation syndrome" during the colonial times when the Whites segregated the Blacks on the pretext that they harbored diseases. In Harare, the COVID-19 mandatory lockdowns were characterized by roadblocks heavily manned by soldiers and police officers from Chitungwiza and other low-income townships to restrict the poor from accessing the CBD; yet this was not the case in the affluent northern suburbs.

Fourth, the book demonstrates poor governance and planning during pandemics and a denial by the authorities. This was the same during the 2008/2009 cholera pandemic and several typhoid outbreaks in Harare. During these times, the government was in denial while people were dying. During the COVID-19 pandemic, the government's response was slow and ineffective, with limited support provided to vulnerable populations in urban areas. The book argues that this denial resulted in inadequate planning and support, which exacerbated the impact of the pandemic on vulnerable populations.

Fifth, the book highlights the impact of top-down planning during pandemics with little consultation or concern about the effects of the decisions on vulnerable urban populations. For example, the lockdown measures implemented in response to the pandemic significantly impacted informal traders and day laborers who could not work during the lockdown period. While one would argue that these top-down measures were necessary and considering the pandemic's complexity, an approach that aligns with muddling through planning process was the best the government could do. The situation may also have been prompted by lack of trust between the government and citizens. This was compromised by the deployment of armed enforcement of lockdown measures, which many believed was too much, considering some individuals reported harassment and brutality from these forces. The book argues that these policies negatively impacted the livelihoods of vulnerable populations and have increased their vulnerability.

Sixth, corruption, embezzlement of funds meant for relief, and blocking aid became apparent during the pandemic. In Zimbabwe, there were reports of funds meant for COVID-19 relief being embezzled by government officials, leading to a lack of support for vulnerable populations. Such practices from public officials depict a gross violation of the rights of the poor, intended to benefit from social grants and aid. What becomes apparent here is the misuse of power at the expense of the poor, who ended up worse off due to the lockdown measures. Further, some organizations providing aid and support were blocked by the government from operating in the country, exacerbating the impact of the pandemic on vulnerable populations. The book calls for more transparency and accountability in using funds for relief and support during pandemics to ensure that vulnerable populations receive the support they need.

Finally, and on a more positive note, the book highlights some positive strides and lessons learnt by the government and authorities in Zimbabwe during the pandemic. The COVID-19 pandemic allowed the government and authorities to reevaluate their policies and strategies for addressing vulnerable populations' needs during crises. Specifically, the pandemic allowed the government and authorities to engage in more collaborative and participatory decision-making processes. For example, there were instances where the government worked with civil society organizations and other stakeholders to address the needs of vulnerable populations during the pandemic. Furthermore, the pandemic led to the development of innovative approaches to addressing the needs of vulnerable urban populations. For example, some organizations implemented digital solutions to support vulnerable urban populations during the pandemic, such as online platforms for accessing healthcare and education. These positive strides and lessons demonstrate the potential for change and progress in addressing the needs of vulnerable urban populations during crises.

Poor Urbanites and the COVID-19 Pandemic

The COVID-19 pandemic affected everyone across the globe, but its impact was particularly harsh on poor urbanites as demonstrated by this book. Urban areas like Harare were the hotspots for the virus, as densely populated cities made social distancing nearly impossible. Poor urbanites in Harare were especially vulnerable as they often lived in crowded conditions with limited access to basic services such as healthcare, sanitation, and clean water. The pandemic exacerbated existing inequalities in urban areas, with the poorest urbanites being hit the hardest. Many of them lost their jobs or experienced reduced income due to the pandemic's economic fallout, making it difficult to afford basic necessities such as food, rent, and healthcare.

Another major challenge poor urbanites faced during the pandemic was access to healthcare. Many lived in areas with limited or no access to healthcare facilities, making it difficult to get tested or receive treatment for COVID-19. Even when healthcare facilities are available, poor urbanites lacked the financial resources to pay for treatment. This led to many cases going untreated, which caused severe consequences for both individuals and the community. Another huge challenge faced by poor urbanites during the pandemic was access to clean water and sanitation. Many live in informal settlements or slums, where access to clean water and sanitation is limited (Bhanye 2022, 2023a; Matamanda 2022). This made it difficult to maintain good hygiene practices, such as frequent handwashing, which is critical to preventing the spread of the virus. The pandemic also had a significant impact on children from poor urban families. Many could not attend school due to closures, which leads to long-term implications for their education and future opportunities. Additionally, many poor urban families relied on school feeding programs to provide their children with regular meals. With schools closed, these families struggled to put food on the table.

Implications for Theory and Urban Practice

The book "COVID-19 Lockdowns and the Urban Poor in Harare, Zimbabwe: Emerging Perspectives and the Morphing of a Sustainable Urban Future" has significant implications for theory and urban practice, particularly in relation to complexity distributive justice, rights to the city, and urban health penalty. In this book, complexity theory was used to understand the behavior of the COVID-19 virus and its impact on different populations, including the urban poor. The urban poor are a complex and diverse group of people who face a range of interconnected challenges, including limited access to healthcare, housing insecurity, and precarious employment. These challenges were amplified by the COVID-19 pandemic, which disrupted supply chains, closed businesses, and limited mobility, making it harder for people to access basic necessities like food and medicine and generate an income. One of the key insights of complexity theory is that small changes in one part of a system can have significant and unpredictable effects on other parts. This means that interventions aimed at addressing the needs of the urban poor during the pandemic need to be holistic and systemic, considering the interconnectedness of different factors that contribute to poverty and vulnerability. Effective interventions during complex crises like the COVID-19 pandemic might include providing targeted financial support to those who have lost income due to the pandemic, increasing access to healthcare and testing in urban areas, and supporting community-based initiatives, which address social and economic needs of the urban poor.

During the COVID-19 pandemic, the issue of distributive justice became particularly important for the urban poor. In this book, distributive justice referred to a situation where resources and benefits were distributed fairly among individuals and groups within a society. The principle of distributive justice holds that people should receive a fair share of benefits and burdens of society based on their needs, abilities, and contributions. However, the urban poor were disproportionately affected by the pandemic in terms of their health and economic well-being. Many urban poor were unable to work due to lockdown and other restrictions and lacked the financial resources to weather the economic downturn caused by the pandemic. Distributive justice theory argues that society should ensure that the burdens of the pandemic are distributed fairly among all members of society, including the urban poor. This may involve providing financial assistance to those unable to work, ensuring access to healthcare and other essential services, and implementing policies to reduce the spread of the virus. Some specific examples of distributive justice measures that could be taken during the COVID-19 pandemic include the following:

1. *Providing financial support to low-income families and individuals who have lost their jobs or had their hours reduced due to the pandemic,*
2. *Ensuring all individuals have access to affordable healthcare and testing, regardless of their income or insurance status,*
3. *Implementing policies to protect essential workers' exposure to the virus, including those in healthcare, grocery stores, and other industries, and*
4. *Providing resources and support to individuals and families struggling with mental health issues related to the pandemic.*

The COVID-19 pandemic highlighted the longstanding issue of the urban health penalty and the unequal access of the urban poor's right to the city. As the virus spread rapidly in cities worldwide, impoverished communities were disproportionately affected due to preexisting health disparities, inadequate access to healthcare, and crowded living conditions. From this book, the urban health penalty referred to the fact that poor urbanites experience worse health outcomes than the elite. This is due to various factors, including exposure to environmental pollution, limited access to healthy food options, and the physical and mental stresses of city living. The pandemic amplified these preexisting health inequalities, as those living in poverty and overcrowded conditions were at higher risk of contracting and suffering from severe cases of COVID-19.

Additionally, the right to the city, as defined by urban theorist Henri Lefebvre as the right for all residents to fully participate in their city's social, political, and economic life, was denied to the urban poor due to systemic inequalities and exclusionary urban planning. The COVID-19 pandemic further exposed this reality. Those living in informal settlements and low-income neighborhoods faced great challenges accessing healthcare, adequate housing, and basic services such as water and sanitation. Despite these challenges, there have been grassroots efforts by community organizations and activists to demand their right to the city and better living conditions in some countries. In some cities, these efforts led to government interventions, such as providing emergency housing and healthcare services.

Morphing a Sustainable, Resilient, Inclusive, and Equitable Urban Future Post-COVID-19 Pandemic

The COVID-19 pandemic disrupted the world unprecedentedly, including how we live, work, and interact with our cities. As the pandemic ends, governments and local authorities can reshape cities into more sustainable, equitable, and resilient spaces, which prioritize the needs of all residents, particularly the most vulnerable. This final chapter explores key strategies for morphing a sustainable and equitable urban future post-COVID-19 pandemic. Investing in sustainable infrastructure is one of the key strategies for creating a sustainable and equitable urban future. This includes developing more efficient public transportation systems, investing in renewable energy sources, and expanding access to green spaces. These investments not only reduce carbon emissions and mitigate the effects of climate change but also promote health and well-being and provide economic opportunities for residents. Another important strategy is prioritizing social inclusion and equity in urban planning and governance. This includes engaging marginalized communities in decision-making, promoting affordable housing and access to basic services such as healthcare and education, and addressing systemic inequalities in urban planning and governance. By prioritizing the needs of all residents, particularly the most vulnerable, we can create more equitable and just cities that promote the well-being of all.

In addition, governments and local authorities must prioritize resilience and preparedness in urban planning. This includes developing strategies for responding to public health emergencies and natural disasters and investing in the resilience of critical infrastructure such as healthcare systems and communication networks. By building resilience into urban planning, governments and local authorities can better prepare for future challenges and ensure that our cities can withstand the shocks and stresses of an increasingly complex and interconnected world. Finally, we must embrace innovation and technology to create a sustainable and equitable urban future. This includes developing smart cities that leverage data and technology to improve public services and enhance the quality of life for residents, as well as promoting innovation in urban agriculture and sustainable food systems. By harnessing the power of innovation and technology, we can create more sustainable and equitable cities that are better equipped to meet the needs of all residents. In summary, the COVID-19 pandemic has given us an opportunity to reshape our cities into more sustainable, equitable, and resilient spaces that prioritize the needs of all residents. By investing in sustainable infrastructure, prioritizing social inclusion and equity, building resilience, and embracing innovation and technology, we can create a brighter and more equitable urban future better equipped to meet the challenges of an increasingly complex and interconnected world.

In Zimbabwe, the context of this book, the COVID-19 pandemic devastated the urban poor, highlighting the preexisting inequalities in the country's cities. However, the pandemic also provided an opportunity to reflect on the current state of urban development in Zimbabwe and to take action to build a more equitable urban future. In this section of the final chapter, we provide recommendations for building a more equitable urban future in Zimbabwe in the post-COVID-19 pandemic phase. These recommendations are based on the insights gained from the preceding chapters and are designed to address challenges faced by the urban poor in Zimbabwe's cities. The recommendations are as follows:

1. *Strengthen social protection programs*: Zimbabwe needs to strengthen its social protection programs. Specifically, these social protection programs may include cash transfers and food assistance which will significantly cover more vulnerable households. In this way, the vulnerability of the urban poor is reduced as the social protection programs provide a safety net for the urban poor during times of crisis and reduce inequality.
2. *Invest in affordable housing and basic services*: Zimbabwe needs to invest in affordable housing and basic services. The central government must commit to engaging the responsible stakeholders and take a lead role in ensuring all citizens have access to services such as clean water and sanitation. However, we recognize and recommend that this will require a comprehensive approach involving both public and private sector investment and partnerships with civil society organizations. This commitment is especially critical in improving the living conditions of the urban poor whose lives and well-being are compromised by the absence and lack of adequate housing and basic services.
3. *Support informal sector workers*: Both the central and local governments in Zimbabwe need to prioritize support for informal sector workers, who have been hit particularly hard by the pandemic due to lockdown measures and reduced economic activity. The negative perception from the government towards the informal sector workers further marginalizes these workers and excludes them from being integrated into the disaster response plans thereby compromising the success of such initiatives. Therefore, the Government of Zimbabwe should consider providing targeted support to informal sector workers, such as access to credit, training, and social protection.

4. *Strengthen healthcare systems*: The Government of Zimbabwe needs to invest in strengthening its healthcare system, including improving access to healthcare services and increasing the availability of medical supplies and equipment. This is critical to redress the inequalities that exist in accessing healthcare facilities. Specifically, mobile clinics can be provided to marginal spaces and individuals. This is especially important considering that the pandemic exposed weaknesses in the country's health system, particularly in terms of capacity and infrastructure.
5. *Build resilience to future shocks*: We recommend that a holistic approach be taken through which the Government Zimbabwe in collaboration with different stakeholders and actors prioritize building resilience to future shocks, such as pandemics and natural disasters. To achieve this, there is a need to invest in early warning systems, disaster preparedness, and emergency response capacities that align with the local realities.

Conclusion

The COVID-19 pandemic significantly impacted urban areas worldwide, exposing and exacerbating preexisting inequalities. Zimbabwe, the City of Harare in particular, was no exception. In this book, we explored the impact of the pandemic and mandatory lockdown policy interventions on the urban poor in Zimbabwe, with chapters focusing on the urban penalty and the right to the city, the socioeconomic context of Harare, policy responses to the pandemic, the impact of COVID-19 on urban inequality, the effects of lockdowns on poor urbanites, and the use of remote ethnography to study COVID-19's effects on urban poverty.

Through these chapters, we gained insights into the challenges faced by the urban poor in Zimbabwe and the need for action to address these challenges. We saw that policy responses to the pandemic were inadequate and that the urban poor were disproportionately affected by the crisis. The urban injustices perpetuated by the COVID-19 pandemic might not be surprising because Africa already has some of the world's most unequal cities (UN-Habitat 2020), with deep-rooted socioeconomic exclusion (Huchzermeyer 2014). African cities inherited colonial segregated planning laws that are socioeconomically exclusive, resulting in cement cities, where the urban poor struggle to achieve the rights to the city (Huchzermeyer 2014). In the decades since independence, few African states have been able to develop and implement reforms governing urban development to effectively improve the right to the city by the urban poor.

Some governments tended to adopt anti-urban policies, fearing the growth of cities and the potential for militancy in the urban underemployed (Collier and Venables 2017). Most of the urban laws and municipal bylaws in place exclude the poor from the benefits of living in a city (Berrisford and McAuslan 2017). With no viable formal alternative, persistently high levels of inequality, and planned segregation in African cities, poor urbanites remake the planned city from below through informal means. However, as demonstrated by our study, COVID-19 pandemic mandatory lockdowns threatened the urban futures of the already marginalized who source food through the informal sector.

As we conclude this book, it is clear that Zimbabwe and other countries in the Global South must take action to build more equitable and sustainable cities in the post-COVID-19 pandemic phase. This will require a comprehensive approach involving policy changes, infrastructure investments, and partnerships with civil society organizations. To build a more equitable urban future in Zimbabwe, we must prioritize the needs of the urban poor, invest in affordable housing and basic services, strengthen social protection programs, support informal sector workers, and build resilience to future shocks. In this book, we recommend that radical policies like mandatory COVID-19 lockdowns should not only be treated as a law-and-order operation but as a medical intervention to stem the spread of the virus backed by measures to safeguard the livelihoods of the urban poor while protecting the economy. This means governments should provide social safety nets to informal sector operators whose income-generating activities are affected the most during emergencies like COVID-19. This may include cushioning allowances and tax breaks for small-scale enterprises, fostering continuity of production in the informal sector in the post-pandemic period to sustain the economy, and making the well-being of vulnerable populations stable.

Beyond immediate relief, COVID-19 lockdown recovery measures need to point towards long-term solutions that contribute to the redistribution of wealth and new urban development by tapping the benefits of more inclusive infrastructure and better access to basic amenities and services for the informal sector. Planners and policymakers should re-envision pandemic-resilient cities that are just, equitable, resilient, and sustainable based on the ideas of the right to the city. This means cities should consider the social function of the city, quality public spaces, and sustainable and inclusive rural–urban linkages; inclusive economies; inclusive citizenship; enhanced political participation; and nondiscrimination, gender equality, and cultural diversity. But beyond romanticizing the right to the city concept as only meaning the urban poor's liberties to access urban space and resources, the approach should encompass the right and ability of individuals to change themselves and utilize their collective power to reshape the processes of urbanization. While these recommendations in this book may not be exhaustive, they provide a starting point for action in morphing a sustainable urban future.

A key research question emerging from the discussions in this book on responses to urban crises and the urban poor is: How do poor urbanites respond when their right to the city is threatened by policies like mandatory lockdowns? While some may comply with the restrictions and others may be relegated to rural areas, as demonstrated in this book, others may likely not remain passive but resist (Bhanye 2023b; Bhanye and Dzingirai 2020). Predictably, there might be more civil disobedience by citizens, radicalization, and increased crime and violence. Other questions that researchers can consider include the following: Do researchers' policy recommendations on the COVID-19 pandemic resonate with government and local government's views? Did the pandemic lead to any changes in how the city is governed or a greater awareness of the crucial issues? Did the authorities understand the dire situation and people's need to continue with associational life in the city during the pandemics?

References

Berrisford S, McAuslan P (2017) Reforming urban laws in Africa: a practical guide. African Centre for Cities, Cities Alliance, United Nations Human Settlements Programme (UN-Habitat) and Urban LandMark, Cape Town, Brussels, Nairobi and Pretoria

Bhanye J (2022) "Lydiate is now our home of a sort": perceptions of place amongst ageing first-generation Malawian migrants in Zimbabwe. Anthropol South Afr 45(3):180–194

Bhanye J (2023a) 'Emerging forms of spatialised and socialised authority' among 'tenure-insecure peri-urbanites' in African peri-urban spaces: a review study. Pan-Afr Convers 1(2):26–63

Bhanye J (2023b) "Emerging forms of Authority in land access?": the occult and witchcraft among Malawian migrants in peri-urban Zimbabwe. In: Debating religion and forced migration entanglements. Springer International Publishing, Cham, pp 89–107

Bhanye J, Dzingirai V (2020) Plural strategies of accessing land among peri-urban squatters. Afr Black Diaspora Int J 13(1):98–113

Collier P, Venables A (2017) Urbanisation in developing economies. Oxf Rev Econ Policy 33(3):355–372

Huchzermeyer M (2014) Humanism, creativity and rights: invoking Henri Lefebvre's right to the city in the tension presented by informal settlements in South Africa today. Transformation 85(2014):64. ISSN 0258-7696

Matamanda AR (2022). Genesis of Hopley Farm informal settlement in independent Harare, Zimbabwe. Afr Geograph Rev 1–14. https://doi.org/10.1080/19376812.2022.2117216

UN-Habitat (2020). UN-Habitat Covid-19: key messages. https://unhabitat.org/sites/default/files/2020/03/Covid19_key_messages_eng_1.pdf. Accessed 14 Apr 2021

Index

J. I. Bhanye et al., *COVID-19 Lockdowns and the Urban Poor in Harare, Zimbabwe*, https://doi.org/10.1007/978-3-031-41669-9

9783031416682